Contents

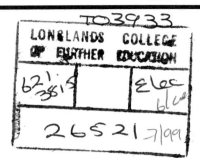

Preface

This book is intended for people with some basic knowledge of electronic principles, who wish to continue to develop an understanding of slightly more complex electronic devices and concepts.

The non-mathematical text commences with decibel notation, which is a sub-plot that runs through electrical and electronic engineering. Familiar analogue electronic topics such as signal amplifiers, power amplifiers and oscillators are discussed in detail. The important, but slightly mysterious concept of electrical noise has been introduced at this level in order to promote an understanding of its real significance in the world of electronics. Feedback is presented in a manner that illustrates its use when designing practical circuits, particularly those using operational amplifiers. Power electronic components like the thyristor and triac are today rapidly replacing electro-mechanical controllers. The operation of these devices together with the design and triggering of circuits is given thorough treatment. An individual, skilled and knowledgeable in the field of electronics should be able to repair faulty equipment. In an effort to be fully comprehensive, the final chapter briefly introduces this subject, showing how basic instruments can be used to diagnose faults in a circuit down to component level.

Discovery-based, student-centred methods are adopted throughout the book that make use of manufacturers' data sheets, self assessment questions and design assignments to reinforce theoretical concepts. In addition to this, hands-on experience can be gained by working through the 24 Practical Investigations. These are designed to help forge the all important link between *knowing* and *doing*. Each chapter includes a review section that allows the salient points to be recapped without excessive *hunting* through the text.

The material covers the principal objectives of the BTEC Level NIII Analogue Electronics Syllabus while sharing common ground with the City and Guilds 224, 271 and GCSE courses. Background information if required can be found in the companion volume *Electronics: Practical Applications and Design* (Morris, 1989). It is my intention that the step-by-step approach to the subject will make this book a source of interest to enthusiasts, technicians and teachers as well as the student of electronics.

The pages within represent the collective efforts of a number of people. I would like, therefore, to acknowledge Farnell Electronic Components Ltd., R.S. Components Ltd. and Maplin Electronics PLC for permission to reproduce extracts from their current catalogues. I shall always be most grateful for the kindness, wit and support of my colleagues, particularly Ben Byrne, Mike Lenard and Chas Taylor whose careful proof reading and helpful suggestions have led to many real improvements.

A project such as this involves a certain family commitment. In this respect Ian and Adam deserve my appreciation for their patience and understanding. To Lin, my wife, I offer a special *thank you* for her skilled help, warm encouragement and masterly typing of the manuscript.

John C. Morris
Billericay, Essex. June 1990

Introduction

The chapters are laid out in a sequence that I consider to be 'natural' for a programmed learning approach. However, if required, each topic can be readily studied in isolation.

In an effort to ensure a thorough understanding I recommend that the associated practical investigations for each topic are carried out. These use proven circuits and can be performed using common components with the minimum amount of normal laboratory equipment. For reasons of safety all a.c. circuits are designed to operate at a low voltage supplied via a mains step-down transformer. Under no circumstances must the specified supply voltage be exceeded. The method of circuit construction used is left to the reader, but the use of a *breadboard* is highly recommended for speed and reusability of components.

Please remember that a 'pioneering spirit' should be evident and that the specified components are only recommendations. If you do not have the appropriate semiconductor — use an equivalent or if a particular resistor is not available — use the nearest value you have. When carrying out the investigations work methodically and pay heed to the following points:

1 Check that the circuit is correct before switching on the supply.
2 Check the supply voltage with a meter when setting to a specified value (meters on power supplies are there for guidance only).
3 Make sure you record everything.
4 Plot graphs *as you go*.
5 *Do not* dismantle your circuit immediately you have finished but check through what you have written and what is yet required so that if you have to repeat part of the procedure the anger and frustration will be minimized.

1

The Decibel and its use

The decibel is a logarithmic ratio between two power levels.

It is usual in the world of electronics to concern ourselves with the gain or attenuation of a circuit, i.e. how much a signal level is increased or decreased as it passes through a circuit or cable. This is often achieved by expressing values of gain or attenuation as ratios, e.g. *in the case of an amplifier* (Fig. 1.1) or *for an attenuating circuit* (Fig. 1.2).

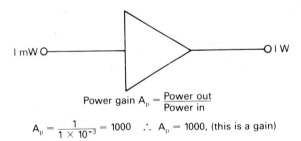

Power gain $A_p = \dfrac{\text{Power out}}{\text{Power in}}$

$A_p = \dfrac{1}{1 \times 10^{-3}} = 1000$ ∴ $A_p = 1000$, (this is a gain)

Fig. 1.1 An amplifying circuit

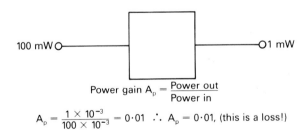

Power gain $A_p = \dfrac{\text{Power out}}{\text{Power in}}$

$A_p = \dfrac{1 \times 10^{-3}}{100 \times 10^{-3}} = 0 \cdot 01$ ∴ $A_p = 0 \cdot 01$, (this is a loss!)

Fig. 1.2 An attenuating circuit

This use of ratios is handy because it tells at a glance exactly what the overall effect of a circuit is; in the case of the amplifier circuit, since it has a power gain of 1000, an input signal emerges

1000 times greater at the output. The attenuating circuit, however, gives an output that is 100 times lower than the input. This concept of a ratio is so informative that there may appear to be little or no reason for using anything other than ratios; but let us consider the practical realities that exist.

Amplifiers can have very large gain values, furthermore they are often connected in cascade (series) to form a complete system. This system

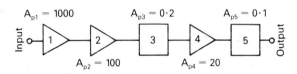

Fig. 1.3 Circuits in cascade

may well include sections that will reduce or attenuate the signal level. We are now in the position of having to calculate the overall effect of a complete circuit. Consider Fig. 1.3 that contains both amplifiers and attenuators. Circuits 1, 2 and 4 are amplifiers, while 3 and 5 are attenuating circuits.

The overall gain

$$(A_p) = A_{p1} \times A_{p2} \times A_{p3} \times A_{p4} \times A_{p5}$$

$$= 1000 \times 100 \times 0.2 \times 20 \times 0.1 = 40\,000$$

The overall result is a power gain $A_p = 40\,000$.

From this some major points emerge:

1 Large and unwieldy values are often involved with circuits having gains of 2×10^9 or losses of 2×10^{-5}.

2   Where circuits are interconnected to form systems, the overall result may involve multiplications of large values, increasing the possibility of error.

3   Often the final result is a number that is so large (or small) that it has little meaning, i.e. we cannot comprehend its *real* value. For example a gain of 40 000 sounds impressive, but what does it really tell us?

To express these numerical ratios in a more practical way, logarithmic ratios can be used. The use of logarithms results in a non-linear or compressed scale. This you may recall if you have ever used logarithmic graph paper in order to fit a large frequency range on a small sheet of paper.

It is also worth noting that many systems respond logarithmically to changes in power levels. The use of logarithmic ratios allows us to linearize this non-linear response. The logarithmic ratio between two power levels is expressed in bels (B), after Alexander Graham Bell.

$$A_p = \text{Log}_{10}\left(\frac{\text{Output power}}{\text{Input power}}\right) \text{ Bels}$$

($\log_{10}$ indicates that logarithms to the base 10 are used).

If $P_{in} = 1\text{ mW}$ and $P_{out} = 1\text{ W}$

$$A_p = \log_{10}\left(\frac{1}{1 \times 10^{-3}}\right) = \text{Log } 1000 = 3\text{ Bels}$$

The bel is a very large unit so for convenience the decibel (abbreviation dB) is commonly used.

1 dB = 0.1 B. (1 dB is one tenth of a Bel!)

This gives the following equation:

$$A_p = 10\,\text{Log}_{10}\left(\frac{P_{out}}{P_{in}}\right)\text{dB}.$$

Note   Once it is understood that logarithms to the base 10 are used the suffix 10 can be omitted.

If we now use the two previous examples, the power gains can be expressed in their dB form.

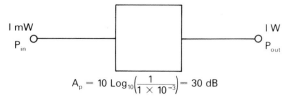

$$A_p = 10\,\text{Log}_{10}\left(\frac{1}{1 \times 10^{-3}}\right) = 30\text{ dB}$$

Fig. 1.4   A power gain of 30 dB

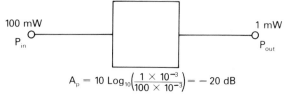

$$A_p = 10\,\text{Log}_{10}\left(\frac{1 \times 10^{-3}}{100 \times 10^{-3}}\right) = -20\text{ dB}$$

Fig. 1.5   A power gain of −20 dB (a loss in fact!)

THOUGHT

So loss or attenuation is always indicated by a minus sign?

Yes *but* be careful! A gain of −x dB is a loss of x dB. You cannot have a loss of −x dB because this would be a double negative!

The use of dBs to express ratios certainly makes the situation much neater. Should there be any doubts about this, a comparison between the numerical gain and decibel gain values will help to dispel them.

A_p (numerical)	A_p in dB
1	0
10	10
100	20
1000	30
1×10^4	40
1×10^6	60
1×10^{10}	100

Use the equation and check these values for yourself.

Circuits in Cascade

If we consider again our original series circuit this time using gain values in dB (Fig. 1.6).

Overall gain $(A_p) = A_{p1} + A_{p2} - A_{p3} + A_{p4} - A_{p5}$
  $= 30\text{ dB} + 20\text{ dB} - 6.98\text{ dB} +$
    $13\text{ dB} - 10\text{ dB}$
  $= 46.02\text{ dB}$

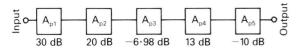

Input o—[A_{p1}]—[A_{p2}]—[A_{p3}]—[A_{p4}]—[A_{p5}]—o Output

30 dB   20 dB   −6·98 dB   13 dB   −10 dB

Fig. 1.6 Circuits in cascade

Overall gain (A_p) = 46.02 dB. As a check let's convert this to a numerical gain value and see how it compares with our original result of 40 000: since A_p in dB = 10 Log (numerical gain).

(Note: *A*.log = *Antilog*.)

$$\text{numerical gain} = A.\log\left(\frac{A_p(\text{in dB})}{10}\right)$$

$$= A.\log\left(\frac{46.02}{10}\right)$$

$$= A.\log 4.602$$

$$= 39\ 994, \text{ i.e. } 40\ 000 - \text{QED!}$$

So where gain values are quoted in dB the overall system gain is simply a matter of adding together all the gains and subtracting the losses.

Change in Signal Level Using Decibels

Another application of the decibel is when a change of signal level needs to be indicated. If a system has an input of 5 mW and an output of 5 mW then clearly neither amplification nor attenuation has taken place. In dB, this is given by:

$$A_p = 10\ \text{Log}\left(\frac{5\ \text{mW}}{5\ \text{mW}}\right) = 0\ \text{dB}$$

So 0 dB indicates no change has occurred!!

Note You must get used to the fact that zero dB does not mean zero gain or zero output, it simply means unity gain.

$$\text{unity} = 1 = \text{No change!}$$

Reference levels

If the input and output powers are known, then a gain or loss in dB can easily be found since the decibel is simply a means of comparing the two power levels; however a reference level is sometimes useful for the following reasons:

1   It gives an indication of the practical realities of a circuit, i.e. it is meaningless to say that an amplifier has a power gain of 60 dB unless some reference power is stated or understood. A power gain of 60 dB represents a numerical gain of 1×10^6 ∴ if the input to such an amplifier was 1 μW the output would be 1 W, fine! But if the input was 1 W the output would be 1 MW − hardly likely! A stated reference level helps to remove any misunderstanding.

2   If the reference level is known, a change in signal level can be readily understood. Once a reference power is used any power level can be considered with respect to this reference level

$$A_p = 10\ \text{Log}\left(\frac{\text{Power level}}{\text{ref. level}}\right)\text{dB}$$

This reference level is indicated by a subscript to the abbreviation for decibel.

dB_m: reference level = 1 mW $(1 \times 10^{-3}\ \text{W})$

$$A_p = 10\ \text{Log}\left(\frac{\text{Power}}{1 \times 10^{-3}}\right)\text{dB}_m$$

dB_w: reference level = 1 W.

$$A_p = 10\ \text{Log}\left(\frac{\text{Power}}{1}\right)\text{dB}_w$$

Example 1
Express in dB_m the following power levels:

(a)   1.5 W   (b)   1 μW   (c)   1 mW

(a) $A_p = 10\ \text{log}\left(\dfrac{1.5}{1 \times 10^{-3}}\right)\text{dB}_m$

$= 32\ \text{dB}_m$ (a gain of 32 dB_m)

(b) $A_p = 10\ \text{log}\left(\dfrac{1 \times 10^{-6}}{1 \times 10^{-3}}\right)\text{dB}_m$

$= -32\ \text{dB}_m$ (a loss of 32 dB_m)

(c) $A_p = 10\ \text{log}\left(\dfrac{1 \times 10^{-3}}{1 \times 10^{-3}}\right)\text{dB}_m$

$= 0\ \text{dB}_m$ (no change)

Self Assessment 1

1   Calculate the power gain in dB$_m$

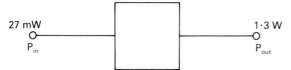

2

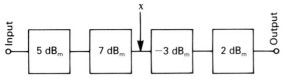

Calculate
(a) The overall system gain in dB$_m$.
(b) The signal power level at point x.
(c) The system output signal power.

Voltage and Current Ratios Expressed in Decibels

So far power levels only have been considered, often however only input and output voltage and current levels are quoted see Fig. 1.7.

Power (P) can be calculated using:

$$P = I^2R \quad or \quad I \times V \quad or \quad \frac{V^2}{R}$$

from this it is easy to see that for any circuit

$$P_{in} = V_{in} \times I_{in} \quad or \quad I_{in}^2 \times R_{in} \quad or \quad \frac{V_{in}^2}{R_{in}}$$

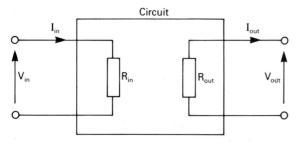

Fig. 1.7   Input and output signals

and

$$P_{out} = V_{out} \times I_{out} \quad or \quad I_{out}^2 \times R_{out} \quad or \quad \frac{V_{out}^2}{R_{out}}$$

Now if the input and output resistances of a circuit are *identical* ($R_{in} = R_{out}$) and if the input and output current or voltage is known, the power ratio can be found using:

$$A_P = 10 \, \text{Log} \left(\frac{(V_{out})^2/R_{out}}{(V_{in})^2/R_{in}} \right) \text{dB}$$

since $R_{in} = R_{out}$ they cancel

To give:

$$A_P = 10 \log \left(\frac{V_{out}}{V_{in}} \right)^2 = 20 \, \text{Log} \left(\frac{V_{out}}{V_{in}} \right) \text{dB}.$$

THOUGHT

So the power gain $A_p = 20 \, Log \left(\dfrac{V_{out}}{V_{in}} \right)$ dB?

True but *only* if R_{in} and R_{out} are identical. Likewise it could be shown that

$$A_p = 20 \log \left(\frac{I_{out}}{I_{in}} \right) \text{dB when } R_{in} = R_{out}$$

Note   Unfortunately R_{in} and R_{out} are seldom identical in real life.

However, it is quite acceptable to express voltage and current ratios in dB, provided it is made clear that they are *not* power ratios. Then:

Power gain $A_p = 10 \log \left(\dfrac{P_{out}}{P_{in}} \right)$ dB

Voltage gain $A_v = 20 \log \left(\dfrac{V_{out}}{V_{in}} \right)$ dB

Current gain $A_i = 20 \log \left(\dfrac{I_{out}}{I_{in}} \right)$ dB

Unfortunately technical literature often quotes a gain in dB without considering R_{in} and R_{out}, we have to put up with this!

Bandwidth

All circuits, transmission lines and cables have a frequency range over which they will perform satisfactorily, this is defined as the bandwidth (B). You may have been used to this being defined as the frequency range over which the current or voltage gain does not fall below 0.707 of its mid-band value. Where numerical ratios are concerned this is absolutely true. However, once decibels are used to express gain the bandwidth is defined as *the frequency range over which the gain falls by no more than 3 dB from its mid-band value*. If the gain/frequency response curve has been plotted using dBs, the bandwidth can be readily defined. This is shown in Fig. 1.8.

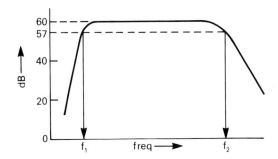

Fig. 1.9   Gain/frequency response

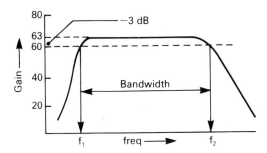

Fig. 1.8   Bandwidth using dBs

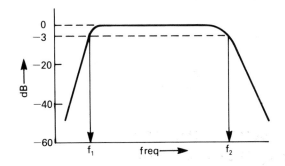

Fig. 1.10   Normalized gain/frequency response

Frequencies f_1 and f_2 are described using a variety of terms, common ones being: the -3 dB frequencies, the -3 dB points, the cut-off frequencies, the break points, the half-power points and the corner frequencies.

Once the frequency response has been plotted it seems reasonable to normalize it so that the mid-band value becomes 0 dB. The bandwidth is then enclosed by its *genuine* -3 dB points. See Figs 1.9 and 1.10.

What do these 3 dB points really mean for power and voltage gain?
Let us consider what a fall in 3 dB means for a power gain.

$$-3\,\text{dB} = 10 \log\left(\frac{P_{\text{out}}}{P_{\text{in}}}\right)$$

$$\therefore \text{As a ratio } A_{\text{p}} = A.\log\left(\frac{-3}{10}\right) = 0.5$$

this represents fall in power gain to 50% of its mid-band value, hence the name *half-power* point.

For a voltage or current gain a loss of 3 dB means:

$$-3\,\text{dB} = 20 \log\left(\frac{V_{\text{out}}}{V_{\text{in}}}\right)$$

$$\therefore \text{As a ratio } A_{\text{v}} = A.\log\left(\frac{-3}{20}\right) = 0.707.$$

representing a fall in voltage gain to 70.7% of its mid-band value.

Sound Intensity

If a survey was carried out the average person would say that the decibel is a measure of volume or level of sound. This is fundamentally untrue because we know that the dB is not an absolute unit, it is simply a logarithmic ratio between two power levels.

However, a sound can be expressed in decibels if a reference level is used.

A typical reference level is a sound so faint it can just be heard by the human ear (in good condition), i.e. a pin dropping or leaves rustling. This is taken to be an audible power level of 1×10^{-12} W/m² and then:

$$\text{Sound intensity} = 10 \log \left(\frac{\text{sound power}}{1 \times 10^{-12}} \right) \text{dB}_\text{A}$$

Where: sound power = noise or sound being measured

1×10^{-12} W/m² = reference power

dB_A = Intensity in dB. The subscript A indicating that the reference level is acoustic at 1×10^{-12} W/m².

THOUGHT

Why W/m² for the reference power level?
This is used because sound intensity decreases as the distance from the source increases and approximately follows the inverse square law, as shown in Fig. 1.11.

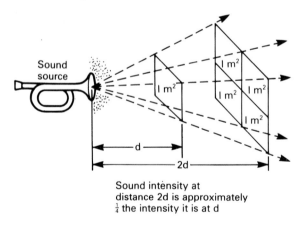

Sound intensity at
distance 2d is approximately
¼ the intensity it is at d

Fig. 1.11   Reduction of sound intensity with distance

We can now quote sound intensity in dB$_\text{A}$ with typical values being:
  Conversation   60 dB
  Noisy factory   85 dB

Car horn at ½   90 dB
Threshold of pain 120 dB — above this level physical pain occurs and ear damage may result.

Note   1 dB is the smallest change in audio power that can be detected by the human ear, i.e. if you set the volume of your audio system to give a particular sound output and then increase the volume until your ears tell you the sound is louder, the output has now increased by 1 dB.

While these values are only approximate they do serve as a guide and indicate clearly the relative intensity levels of the environment in which we live. For audio equipment the dB is widely used, with most frequency response curves drawn to show maximum or mid-band gain as 0 dB. *VU meters* (Volume Units) show the maximum power output into loudspeakers or magnetic tape as 0 dB. Even volume controls have been renamed *gain* controls, having maximum output marked as 0 dB and everything below this as −dB. I suppose this almost universal use of the decibel should lead to a widespread understanding of its meaning, but alas I fear that for the most part it passes unnoticed; you, however, will become very familiar with the dB since it is also used for the measurement of signal-to-noise ratios in Chapter 6.

Decibel Review

1   The Decibel (dB) is not an absolute unit but the logarithmic ratio between two power levels.

2   Power gain $A_\text{p} = 10 \log_{10} \left(\dfrac{P_\text{out}}{P_\text{in}} \right)$ dB.

(Suffix 10 indicates logarithms to the base 10 are used, its inclusion in the equation is optional)

3   A gain is indicated by positive dB values, a loss by negative dB values.

4   Voltage gain $A_\text{v} = 20 \log \left(\dfrac{V_\text{out}}{V_\text{in}} \right)$ dB.

5   Current gain $A_\text{i} = 20 \log \left(\dfrac{I_\text{out}}{I_\text{in}} \right)$ dB.

6 The bandwidth of a system is determined by the upper and lower frequencies at which the gain has fallen by 3 dB from its mid-band value.
7 −3 dB represents a fall to 0.707 of the mid-band voltage and current gain.
8 −3 dB represents a fall to 0.5 (half-power) of the mid-band power gain.
9 Reference levels can be used indicating the power that any gain or loss is compared to. The actual reference power level is expressed using a subscript letter:
$dB_m = 1$ mW $dB_w = 1$ watt
$dB_A = 1 \times 10^{-12}$ W.
(remember subscript A = acoustic)
10 For circuits in cascade the overall effect of the system can be found by adding (algebraically) the gains and losses.
11 It is usual to normalize a frequency response curve so that maximum output or mid-band gain is represented by 0 dB.
12 Sound intensity can be quoted in dB using:

$$intensity = 10 \log \left(\frac{Power}{1 \times 10^{-12}} \right) dB_A$$

13 1 dB is the smallest change in audio power that can be detected by the human ear.

Self Assessment Answers

Self Assessment 1

1 $A_p = 10 \log \left(\dfrac{P_{out}}{P_{in}} \right)$ dB

$= 10 \log \left(\dfrac{1.3}{27 \times 10^{-3}} \right) dB = 16.8$ dB

(a) Overall system gain $= 5 + 7 - 3 + 2$ dB_m
    $= 11$ dB_m
(b) The signal power at x $= 12$ dB_m gain on input signal
    input signal $= 1$ mW (dB_m implies a reference signal of 1 mW)

    numerical value of gain for 12 dB

    $= A \log \left(\dfrac{12}{10} \right) = 15.85$

    output at x $= 1 \times 10^{-3} \times 15.85 = 15.85$ mW.
(c) Output signal level $= 11$ dB_m gain on input signal
    input signal $= 1$ mW

    numerical value of 11 dB

    $= A . \log \left(\dfrac{11}{10} \right) = 12.59$

    output signal $= 1 \times 10^{-3} \times 12.59 = 12.59$ mW.

2
Amplifiers

Amplifiers can really be divided into two types:

1 The small signal amplifier.
2 The power amplifier.

The small signal amplifier is used to increase a very low level voltage or current signal to a level that can be more easily handled. It must do this without introducing much distortion or noise. For this reason very small changes in voltage and current are employed in the amplifying device itself. Hence the name *small signal*. These amplifiers are often called pre-amplifiers because they boost very tiny signals to a level that can be used to drive a power amplifier.

The power amplifier is invariably a 'beast' that takes the output from a pre-amplifier and boosts it to such a level that it can do physical work. Examples of this include driving the cone of a loudspeaker in and out to produce an audible sound output, delivering power to a motor that is part of a control system, or providing radio frequency (rf) power to an aerial so that a signal may be transmitted. When considering such applications it is easy to see that the requirement of a power amplifier may be anything from a couple of watts to many kilowatts. Consequently very large changes in current and voltage occur in the amplifying device, hence the alternative name of *large signal amplifier*.

Transistor Biasing

From previous encounters you may have had with electronics you will be aware that perhaps the most popular circuits are the common emitter and its FET equivalent the common source amplifier, shown in Figs 2.1 and 2.2.

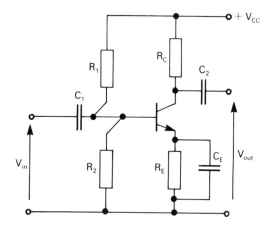

Fig. 2.1   Common emitter amplifier

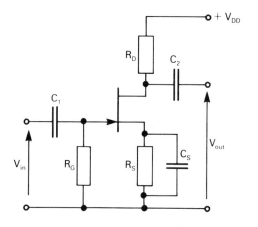

Fig. 2.2   Common source amplifier

You may also know that such amplifiers are usually operated in Class A bias. We need now to examine more closely what this term biasing means.

Biasing a transistor means establishing potential differences across its electrodes. These will cause currents to flow in the device. Voltages and currents set up by the biasing establish the d.c. or static conditions, referred to as the *quiescent* or *quiet* state: so called because there is no input signal connected to the amplifier. The transistor itself is at *quiescence*, it is d.c. biased and so has voltages and currents present but is not actually amplifying any input signals. Obviously the size of the d.c. bias voltages will determine the quiescent current flowing in the device. To help determine the actual bias points required under quiescent conditions a load line is drawn on the output characteristics as shown in Figs 2.3 and 2.4.

From this it appears that the transistor could be biased at *cut-off* (point x) or *saturation* (point y) or anywhere in between these two points.

Once the bias point has been fixed it becomes known as the *quiescent* or Q point. If an input signal is now applied it will drive the Q point up and down the load line as it swings positive and negative, this in turn will produce changes in the output current and voltage. The position chosen for the Q point on the load line is critical for the following reasons:

1   It will determine the maximum possible change in output current and voltage, i.e. if the Q point is fixed near the top of the line any input signal will not be able to drive the point very far up the line before the transistor saturates. Likewise, a bias point near the bottom will mean that when the input signal swings negative the Q point will not have far to move before the transistor is *cut off*.

2   The Q point determines the quiescent current that flows in the device. Consequently, if the transistor is biased near the top of the line, the d.c. current flowing in the device will be high and this means that power is being consumed even though there is no input signal to the amplifier. Conversely a Q point near the bottom of the line will ensure that a low d.c. current will flow under quiescent conditions.

It is the d.c. biasing that determines the Q point,

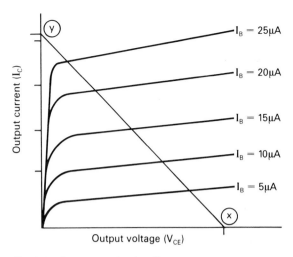

Fig. 2.3   Common emitter load line

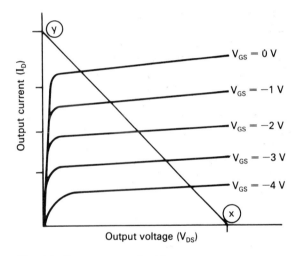

Fig. 2.4   Common source load line

the Q point that determines the *class* of amplifier and the amount of distortion produced.

Class A Biasing

From a study of the output characteristics of Figs 2.3 and 2.4 the best place to bias the device appears to be so that the Q point is in the middle of the line. This is Class A biasing and offers the advantage that maximum changes in output current and voltage are possible because the Q point can move an equal amount up and down the line as indicated in Fig. 2.5(a). Let us consider the voltage and current changes that occur when an input signal is applied. Study Figs 2.3, 2.4 and 2.5(a) in conjunction with the tables below.

Table 1.1   Common emitter amplifier (using an n-p-n transistor)

Input Signal Swings positive	Input Signal Swings negative
I_B increases I_C increases V_{CE} decreases	I_B decreases I_C decreases V_{CE} increases

Table 1.2   Common source amplifier (using an n channel transistor)

Input Signal Swings positive	Input Signal Swings negative
V_{GS} becomes less negative I_D increases V_{DS} decreases	V_{GS} becomes more negative I_D decreases V_{DS} increases

A further advantage of Class A bias is that the transistor will be operating over the most linear region of its characteristic. This cannot be seen from the output characteristic but is quite obvious from the input characteristic for a BJT (Fig. 2.5(b)) and the transfer characteristics of an FET (Fig. 2.5(c)).

It would appear that Class A bias is the only sensible one from all standpoints. There are, however, other considerations:

(a) Supposing the signal to be amplified was one polarity only, i.e. a negative or positive pulse — Class A bias would not then allow the maximum input signal to be applied since the device is already half turned on (or off according to your point of view).

(b) Bias current is flowing in the transistor even when there is *no* input signal!! Consider Fig. 2.5(a) without an input signal applied. The quiescent current flowing is 5 mA; this means that the Class A amplifier is consuming power doing nothing — hardly efficient!

When considering an input signal, a device is said to be biased in Class A when output current

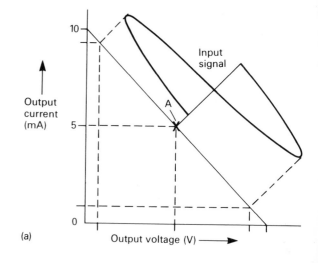

(a)

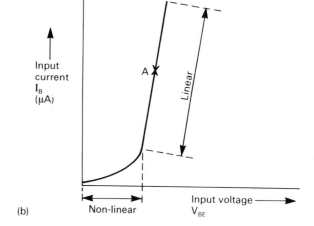

(b)

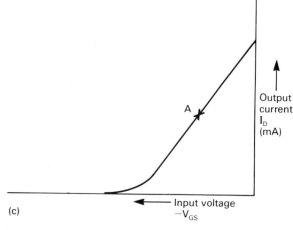

(c)

Fig. 2.5   (a) Class A bias; (b) BJT input characteristic; (c) FET transfer characteristic

flows for a full cycle (360°) of a suitable sinusoidal input signal.

THOUGHT

So common emitter and common source amplifiers are always biased in Class A and you can tell this from the circuit diagram layout? — NO! The biasing is determined by the actual value of the bias resistors. You cannot say from the appearance of this circuit the type of bias employed.

Class A amplifiers are inefficient, having a theoretical efficiency of 50% (at full power) but with the practical efficiency closer to 25%. This makes them unsuitable for power amplification but very suitable for pre-amplifier use, where they offer a very high gain with low distortion of the output signal.

Class B Biasing

Here the amplifying device is actually biased at cut-off (Fig. 2.6). The biasing components ensure that the transistor is not conducting *at all* under quiescent conditions. This means that no d.c. current will flow. The input signal will turn the transistor *on* and only then will current flow in the load. You will see by examining Fig. 2.6(a) that if a sine wave is applied, current will only flow in the load for one half of the input cycle (180°). In this case the positive half (since the device is already off, the negative half will only serve to turn it off more!).

This makes the Class B amplifier more efficient than the Class A since when there is no input signal applied there is no current flowing in the output and hence no power is consumed.

However, when the transistor is turned *on* the base-emitter junction voltage (V_{BE}) must be overcome. This can be considered as having the same characteristic as a forward biased diode, meaning that the non-linear part must be overcome during turn-on, thus causing distortion (Figs 2.6(b) and 2.6(c)). This will be covered in greater detail when power amplifiers are considered.

Class AB Biasing

Clearly Class A and Class B are the two extremes. It is often preferable to bias a transistor so that

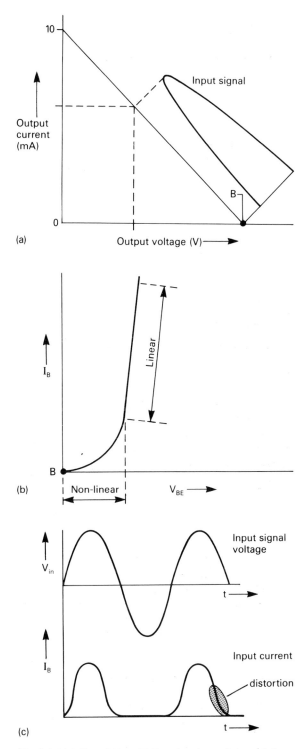

Fig. 2.6   (a) Class B bias; (b) Transistor input characteristic; (c) Distortion produced by non-linear region

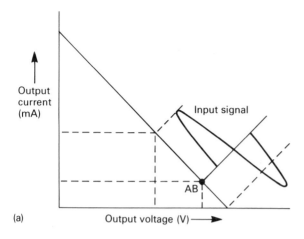

(a)

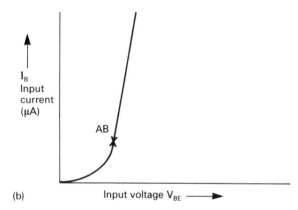

(b)

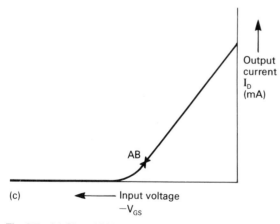

(c)

Fig. 2.7   (a) Class AB bias; (b) BJT input charactereistic; (c) FET transfer characteristic

under quiescent conditions it is not completely *cut off* nor is it in Class A — but somewhere between the two, as shown in Fig. 2.7(a), (b) and (c). This means that current will flow in the output for more than 180° of the input signal but less than 360°, thus making the Class AB amplifier more efficient than a Class A but less efficient than a Class B.

Now perform the investigation on biasing and discover for yourself how the amplifiers performance is determined by the class of bias that is used.

Class C Biasing

This is quite special for here the device is biased beyond its cut-off point (Fig. 2.8(a)) i.e. it is held *hard-off*. The input signal must overcome this *turn-off* bias and then turn the device on in order to produce current flow in the load. The result of this is that the device will conduct for less than one half cycle (less than 180°) giving an output that may resemble a series of blips. This makes the Class C amplifier very efficient (approximately 100%) but causes considerable distortion — a fact that is unimportant where tuned amplifiers are concerned — more about this later.

Comparison of Class A, B, AB and C amplifiers

Let us now compare the different classes of biasing in order to develop an appreciation of their relative advantages and disadvantages. To help visualize the bias points, all classes are shown on the input characteristic of a bipolar transistor in Fig. 2.9(a) and on the transfer characteristic of a field effect transistor in Fig. 2.9(b).

Class A

Biased so that under quiescent conditions the Q point is approximately midway along the load line ∴ current flows in the output for a full 360° of an input sine wave.

Advantages — A very good quality undistorted output signal.
Disadvantages — Current flows in the load even when there is no input signal, this makes the Class A amp only about 25% efficient in practice.

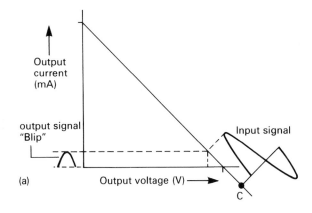

(a)

Output current (mA)

output signal "Blip"

Input signal

Output voltage (V) →

C

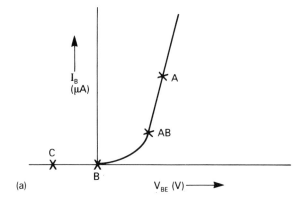

(a)

I_B (μA)

A

AB

C

B

V_{BE} (V) →

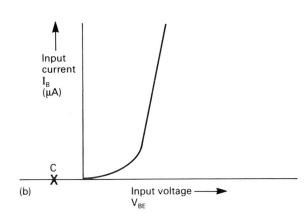

(b)

Input current I_B (μA)

C

Input voltage V_{BE} →

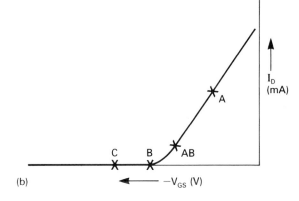

(b)

I_D (mA)

A

C   B   AB

← $-V_{GS}$ (V)

Fig. 2.9   Classes of biasing shown in (a) BJT input characteristic; (b) FET transfer characteristic

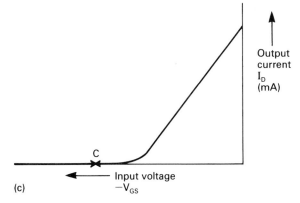

(c)

Output current I_D (mA)

C

Input voltage $-V_{GS}$ →

Fig. 2.8   (a) Class C bias; (b) BJT input characteristic; (c) FET transfer characteristic

Application — pre-amplifiers and low power audio amplifiers, instrument amplifiers.

Class B

Biased so that the Q point is at cut-off, so current flows in the output for 180° of input signal.

Advantages — Improved efficiency due to the fact that no power is consumed under quiescent conditions, this gives a practical efficiency of 50%.

Disadvantages — Due to the non-linear part of the transistor's input characteristic, current will not flow until V_{BE} reaches about 0.6 V; this causes distortion.

Application — Audio power amplifiers, d.c. servo motor amplifiers and P.A. amplifiers. To achieve this without distortion two amplifying devices are used (see Chapter 3 page 44).

Classes of Transistor Biasing

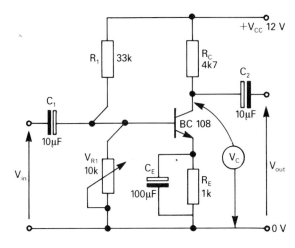

Equipment
CRO
Signal generator
Digital voltmeter
Power supply
General-purpose transistor,
e.g. BC108
33 k, 4k7, 1 k resistors
10 k variable resistor
2 × 10 μF, 1 × 100 μF capacitors

Method
1   Build the circuit as shown.
2   Monitor V_C with the digital voltmeter and adjust V_{R1} until V_C is approximately 6 V $\left(\dfrac{V_{CC}}{2}\right)$
    The transistor is now biased in Class A.
3   Connect the signal generator to provide V_{in} and adjust to give a 1 kHz, 30 mV peak-to-peak
    sine wave.
4   Monitor V_{in} and V_{out} using the CRO and sketch these on a common time scale.
5   By adjusting the level of V_{in} determine the signal distortion level (the maximum value of V_{in}
    before V_{out} distorts!).
6   Observe the affect of varying the bias level V_{R1}.
7   Disconnect the signal generator.
8   Monitor V_C with the digital voltmeter and adjust V_{R1} until V_C is 11.95 V the transistor is now
    biased in Class B.
9   Reconnect the signal generator and adjust to give a V_{in} of 1 kHz 800 mV peak-to-peak.
10 Sketch V_{in} and V_{out}.
11 Observe the effect of varying V_{R1}.

Results
Compare the waveform sketches of both tests.
1   Which class of bias gives an undistorted output?
2   Which class of bias is the most economical and why?

Class AB

Biased between Class B and Class A, current flows in the output for more than 180° but less than 360° of input signal.

Advantages — Part of the distortion caused by Class B biasing is overcome. This improves the quality of the output signal.
Disadvantages — Less efficient than Class B. Actual efficiency depends upon the position of the Q point.
Application — High-quality audio power amplifiers.

Class C

Biased beyond cut-off ∴ output current will flow for *less* than 180° of the input signal.

Advantages — Very efficient with figures in excess of 75% possible.
Disadvantages — Incredible distortion produced.
Application — Tuned R.F. power amplifiers, oscillators.

More About Load Lines

The d.c. load line enables us to choose a quiescent bias point for our transistor. As the name suggests it defines the current and voltage levels that exist under no signal or static conditions. When an alternating signal is applied the d.c. signals will vary and the device will be under operating or a.c. conditions. The circuit will now perform slightly differently since any capacitors used for coupling or decoupling purposes can be considered as short circuits to a.c. signals. We can therefore draw an a.c. load line as well as the d.c. one.

The d.c. load line

Consider the circuit shown in Fig. 2.10 of the common emitter amplifier and the transistor's output characteristic (Fig. 2.11).

To construct the load line two points only are required:

point $x = I_C$, when $V_{CE} = 0$ V.
point $y = V_{CE}$, when $I_C = 0$.

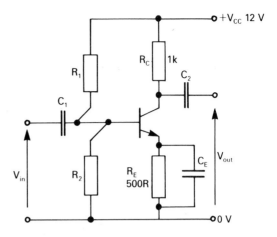

Fig. 2.10 Common emitter amplifier

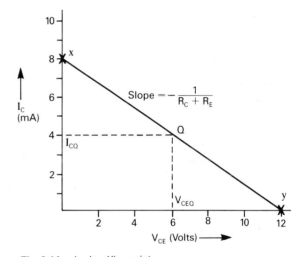

Fig. 2.11 d.c. load line points

point y When $I_C = 0$ $V_{CE} \simeq V_{CC}$ (since no volt drop will occur across R_C). V_{CE} (max) = 12 V = point y.
point x When $V_{CE} = 0$ V maximum I_C will flow.

$$\therefore I_C = \frac{V_{CC}}{R_C + R_E} = \frac{12}{1500} = 8 \text{ mA}$$

$$\therefore I_C \text{ (max)} = 8 \text{ mA} = \text{point } x$$

Showing that the maximum d.c. load on the transistor is 8 mA. By joining these two points the d.c. load line can be drawn and a quiescent point selected, i.e. approximately half way if Class A bias is desired, giving

$$V_{CEQ} = 6 \text{ V} \qquad I_{CQ} = 4 \text{ mA}.$$

(*Note* the use of subscript Q to indicate quiescent values.) This line has a slope of $-1/R_C + R_E$ the minus indicating that an increase of 1 volt in output voltage (V_{CE}) will produce a *reduction* in output current (I_C) of

$$\frac{-1}{1500} = -0.66 \text{ mA/V}$$

THOUGHT

So if V_{CE} changes by 12 V, I_C changes by 12 × 0.66 mA = 8 mA? — Correct.

The a.c. load line

Under signal conditions all capacitors can be considered to be short circuit, so the a.c. load on the transistor will be different to that under d.c. conditions. C_E bypasses R_E so the maximum collector current (I_C) will now be given by:

$$I_C = \frac{V_{CC}}{R_C}$$

This will have a slope of

$$\frac{-1}{R_C} = \frac{-1}{1000} = -1 \text{ mA/V}$$

Indicating that for every increase in collector current (I_C) of 1 mA, V_{CE} will fall by 1 volt; remember that under d.c. conditions a 0.66 mA rise in collector current caused a fall in V_{CE} of 1 volt.

To draw the a.c. load line

Once the slope of the line has been determined the a.c. load line can be drawn using the existing Q point and one other (Fig. 2.12).

The a.c. input signal will cause I_C and V_{CE} to change so:

1   Decide on a change in V_{CE} that will not drive it to extremes, e.g. ±2 V.
2   Calculate the change in I_C that will produce a 2 V change in V_{CE} using

$$2 \times \frac{1}{R_C} = 2 \times 1 \text{ mA} = 2 \text{ mA}.$$

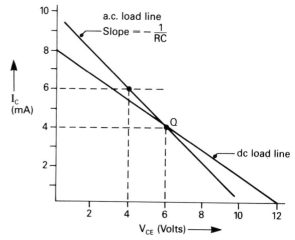

Fig. 2.12   a.c. load line construction

3   Determine the value of I_C when V_{CE} has fallen by 2 V. As V_{CE} falls by 2 V so I_C increases by 2 mA, when $V_{CE} = 4 \text{ V}$
$I_C = 4 \text{ mA} + 2 \text{ mA} = 6 \text{ mA}$.
4   Mark the spot where the new value of I_C and V_{CE} intersect. $I_C = 6 \text{ mA}$ $V_{CE} = 4 \text{ V}$.
5   Draw a line through this point and the Q point to give the a.c. load line. (There is no need to project this line to meet the horizontal and vertical axis.)

This shows clearly that under operating conditions the amplifier performs slightly differently due to the behaviour of the reactive components; you must remember also that whatever is connected to the amplifier will act in parallel with the load resistor under signal conditions and will thus alter the slope of the load line.

Self Assessment 2

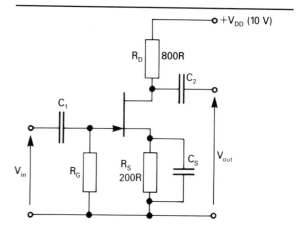

If the amplifier shown is biased in Class A; sketch the output characteristics together with the d.c. and a.c. load lines showing how the construction points have been calculated.

Transistor Small Signal Equivalent Circuits

It is now apparent that a transistor is biased using a d.c. supply to establish the quiescent or static conditions. When a signal is applied these carefully chosen values will vary according to the input signal fluctuations. So for a common emitter amplifier the biasing determines static values of:

- Base current (I_B)
- Collector current (I_C)
- Base-Emitter Voltage (V_{BE})
- Collector-Emitter Voltage (V_{CE})

Likewise for a common source FET amplifier the static values are:

- Gate-source voltage (V_{GS})
- Drain current (I_D)
- Drain-source voltage (V_{DS})

Notice that upper case symbols and subscripts are used to denote that the values quoted are the d.c. or static ones.

When a signal is applied these static values will increase and decrease about their quiescent value as the input signal swings positive and negative, i.e. if an amplifier has $I_{BQ} = 20\,\mu A$, $I_{CQ} = 3\,mA$ and $V_{CEQ} = 5.0\,V$ (Q indicates quiescent values). When an input signal is connected I_B varies by $\pm5\,\mu A$, I_C varies by $\pm1\,mA$ and V_{CE} varies by $\pm2.0\,V$.

We now have an a.c. signal component present that is changing the static values. I_B swings from $20\,\mu A$ to $25\,\mu A$ and back to $15\,\mu A$, I_C changes from $3\,mA$ up to $4\,mA$ and down to $2\,mA$, while V_{CE} varies from $5.0\,V$ up to $7.0\,V$ and down to $3.0\,V$.

These small changes are expressed using the Greek letter delta (δ).

So $\delta I_B = 10\,\mu A$ $\delta I_C = 2\,mA$ $\delta V_{CE} = 4.0\,V$

To show that a current or voltage is a.c., lower case subscripts are used:

$\therefore I_b = 10\,\mu A$ $I_c = 2\,mA$ $V_{ce} = 4.0\,V$.

Where the lower case subscripts indicate small signal rms values

$\therefore \delta V_{GS} = V_{gs}$ $\delta V_{DS} = V_{ds}$

$\delta I_D = I_d$ for a common source amplifier.
This may seem confusing so let's recap:

Upper case symbol and subscript (I_B, V_{BE}, V_{GS} etc.) indicates d.c. or static values only.
Upper case symbol and lower case subscript (I_b, V_{be}, V_{gs} etc.) indicate small signal rms values.

The reason for labouring this aspect of transistors is that the devices are mostly used for operation with varying signals as amplifiers, and so manufacturers tend to quote typical values under signal conditions. To design or understand a circuit it helps to be able to predict the performance of a device under operating conditions. To do this a circuit model can be drawn that considers only the small signal or a.c. operation. Before we can do this the manufacturers' quoted parameters for the BJT and FET must be examined.

The BJT small signal 'h' parameters

These are used to define the operation of a transistor under small signal conditions; remember *small signal* means that any signal change is occurring over the linear region of the transistor's characteristics. The symbol h stands for *hybrid* or mixture — so called because the parameters consist of a mixture of current and voltage ratios, input resistance and output conductance. To help in understanding what this means it is perhaps best to consider the transistor as a *black box* with 2 input and 2 output terminals (often referred to as a two port network), shown in Fig. 2.13(a).

Lower case subscripts are used to indicate that the quantities are small signal changes in voltages and currents.

I_i = small signal input current
I_o = small signal output current
V_i = small signal input voltage
V_o = small signal output voltage

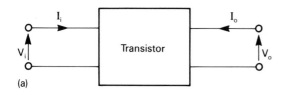

(a)

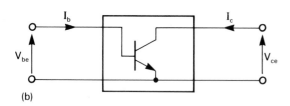

(b)

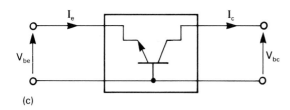

(c)

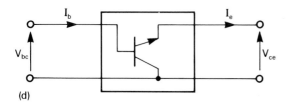

(d)

Fig. 2.13 (a) Transistor as a two port network; (b) Common emitter mode; (c) Common base mode; (d) Common collector mode

There are four h parameters related to the BJT and these are:

h_i = small signal input resistance in ohms
h_f = small signal forward current transfer
    ratio
h_r = small signal reverse voltage ratio
h_o = small signal output conductance in
    Siemens

With reference to the diagram of Fig. 2.13(a) the h parameters can be defined with respect to the input and output signals.

$$h_i = \left(\frac{V_i}{I_i}\right) \text{ When } V_o = 0 \text{ (ohms)}$$

$V_o = 0$ means that the output voltage is held constant at its d.c. bias level.

$$h_f = \left(\frac{I_o}{I_i}\right) \text{ When } V_o = 0$$

(ratio no units)

$$h_r = \left(\frac{V_i}{V_o}\right) \text{ When } I_i = 0$$

(ratio no units) $I_i = 0$ means that the input current is constant at its d.c. bias level.

$$h_o = \left(\frac{I_o}{V_o}\right) \text{ When } I_i = 0$$

Siemens (A/V)

Note   If you are bothered by the idea of one term being zero consider Fig. 2.13(a) again. There are four quantities that can vary V_i, I_i, V_o and I_o. To determine the operation of such a network it makes life simple if we hold one of the variables steady, i.e. unchanging from a signal point of view. Two of the quantities can now be changed while the remaining one is observed.

A transistor can be connected in three possible modes: common emitter, common base and common collector. We can now use the two port network and have real terms for the input and output signals as shown in Figs 2.13(b), (c) and (d). We can now define the h parameter for each configuration; to indicate which mode of connection is being considered a second subscript is added, e.g. h_i is the input resistance so h_{ie}, h_{ib}, h_{ic} is the input resistance for common emitter, common base and common collector connections, respectively.

THOUGHT

So the first letter h indicates hybrid parameter, the first subscript indicates the term that is being considered, i.e. i = input, f = forward etc., and the second subscript indicates the mode of connection? This is correct and remember lower case letters indicate small signals are involved.

The common emitter configuration is a popular way of connecting a transistor as an amplifier so let us now see exactly how the *h* parameters for this are derived.

h_{fe} = small signal forward current gain

$$= \left(\frac{I_c}{I_b}\right) \text{ When } V_{ce} = 0 \text{ (ratio } no\ units!)$$

($V_{ce} = 0$ means that the output voltage is held constant at its d.c. bias level.)

h_{ie} = small signal input resistance

$$= \left(\frac{V_{be}}{I_b}\right) \text{ When } V_{ce} = 0 \text{ (ohms)}$$

($V_{ce} = 0$ means that the output voltage is held constant at its d.c. bias level.)

h_{re} = small signal reverse voltage ratio

$$= \left(\frac{V_{be}}{V_{ce}}\right) \text{ When } I_b = 0 \text{ (ratio: no units)}$$

($I_b = 0$ means that the input current is held constant at its d.c. bias level.)

h_{oe} = small signal output conductance

$$= \left(\frac{I_c}{V_{ce}}\right) \text{ When } I_b = 0 \text{ (in siemens (S))}$$

($I_b = 0$ means that the input current is held constant at its d.c. bias level.)

h parameter values for the common base and common collector configurations can be determined using the same technique. Examine Table 2.1 to get an idea of typical values for the three modes of connection of a general purpose Si transistor.

Table 2.1 Typical h parameter Values for a general purpose Si transistor.

Common Emitter	Common Base	Common Collector
$h_{fe} = 100$	$h_{fb} = 0.99$	$h_{fc} = 101$
$h_{ie} = 5\,k\Omega$	$h_{ib} = 50\,\Omega$	$h_{ic} = 5\,k\Omega$
$h_{re} = 10 \times 10^{-4}$	$h_{rb} = 10 \times 10^{-4}$	$h_{rc} = 1$
$h_{oe} = 25\,\mu S$	$h_{ob} = 1\,\mu S$	$h_{oc} = 25\,\mu S$

Field effect transistor small signal parameters

The FET is a voltage-operated device with a very high input resistance, which leads to fewer parameters for consideration under small signal conditions.

gm = the mutual conductance

$$= \left(\frac{I_d}{V_{gs}}\right) \text{ siemens (S)}$$

(sometimes called *gfs* or *yfs*).

rds = resistance between drain and source

$$= \left(\frac{V_{ds}}{I_d}\right) \text{ ohms}$$

With these facts now established let's do some modelling — starting with the FET because it is the simplest.

The small signal FET equivalent circuit

The FET is a voltage controlled device.

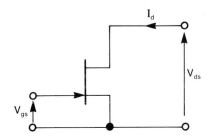

Fig. 2.14 JFET a.c. signals

The transistor itself can be considered as a current generator (symbol ⊖) that is controlled by an input voltage signal.

Note The FET has such a high input resistance that the gate is shown open circuit.

So the circuit symbol for the FET can now be replaced by this equivalent electrical circuit with the real parameters indicated (Figs 2.15 and 2.16).

Rules for drawing equivalent circuits

Any circuit can be redrawn as a small signal model provided that two basic rules are followed:

1 *All* capacitors are considered short circuit to small signals due to their low reactance and

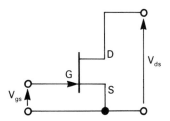

Fig. 2.15 JFET in common source mode

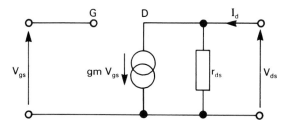

Fig. 2.16 a.c. equivalent circuit

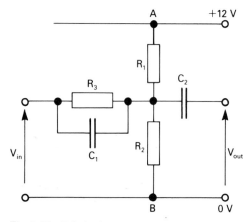

Fig. 2.17 Full circuit

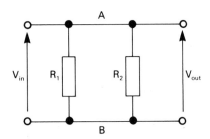

Fig. 2.18 a.c. equivalent circuit

consequently they do not appear in the equivalent circuit. If this is the case, any component shunted by a capacitor is also not drawn since it will be *bypassed* under signal conditions!

2 d.c. power supplies or batteries act like capacitors to signal currents and so can also be considered short circuit under signal conditions.

As an example, circuit Fig. 2.17 could be redrawn under signal conditions as the equivalent circuit Fig. 2.18.

C_1 and C_2 are short circuit to a.c. signals so R_3 is not shown. The d.c. power supply is also short circuit to a.c. so point A is connected to point B. R_1 can thus be considered to be in parallel with R_2 from a signal point of view.

Make sure you understand what is happening here before moving on.

The Common Source Amplifier

The circuit of Fig. 2.19 can be redrawn as the a.c. equivalent circuit shown in Fig. 2.20.

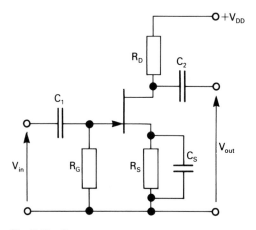

Fig. 2.19 Common source amp.

The voltage (A_v) gain of this amplifier can be calculated using $A_v = -gm \times rds \,\|\, R_D$ (*remember $\|$ means in parallel*).

$$A_v = -gm \left(\frac{rds + R_D}{rds \times R_D} \right)$$

(*remember the minus sign indicates phase inversion*)

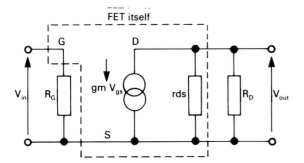

Fig. 2.20 a.c. equivalent circuit of a common source amp.

Since *rds* is usually much greater than R_D the voltage gain approximates to

$$A_v = -gmR_D$$

THOUGHT

Does this mean that the gain A_V can be determined by the drain resistor R_D?
— Yes.

Input Resistance R_{in}

The FET itself has a very high input resistance so the input resistance of the amplifier will be determined by R_G.

$$\therefore \; R_{in} \simeq R_G$$

Output Resistance R_{out}

This will be approximately equal to *rds* in parallel with R_D

$$R_{out} \simeq rds \| R_D$$

($\|$ means in parallel with) so $R_{out} \simeq \dfrac{rds \times R_D}{rds + R_D}$

The BJT equivalent circuit

The BJT is a current operated device where I_b controls I_c which in turn determines V_{ce}.

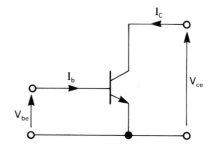

Fig. 2.21 BJT a.c. signals

The transistor can be considered as a current generator controlled by the current flowing in an input resistance (h_{ie}). See Figs 2.22 and 2.23.

Note that the current generator is shunted by the reciprocal of the output conductance since

$$\frac{1}{\text{conductance}} = \text{resistance,}$$

e.g. if $h_{oe} = 30\,\mu S$, $\quad \dfrac{1}{h_{oe}} = 33.3\,k\Omega$.

Now Consider the Common Emitter Amplifier

See Fig. 2.24 and its equivalent circuit Fig. 2.25.

From this R_{in}, A_i and A_v can be estimated.

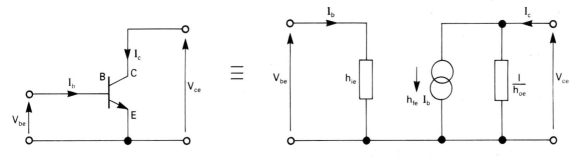

Fig. 2.22 BJT in common emitter mode

Fig. 2.23 a.c. equivalent circuit

22 *Amplifiers*

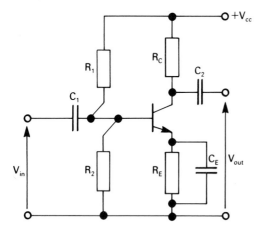

Fig. 2.24 Common emitter amp.

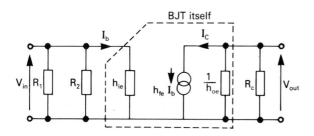

Fig. 2.25 a.c. equivalent circuit

R_{in} This can be approximated as R_1, R_2 and h_{ie} in parallel.

$\therefore R_{in} = R_1 \| R_2 \| h_{ie}$, but since h_{ie} is usually much lower than the other two $R_{in} \simeq h_{ie}$.

Current gain A_i

$$A_i = \frac{I_c}{I_b} \ remember \ I_c = h_{fe} I_b$$

(when $V_{ce} = 0$ V)

$$\therefore A_i = \frac{h_{fe} I_b}{I_b} \simeq h_{fe}$$

Output Resistance R_{out}

Now from the equivalent circuit it appears that R_{out} is the parallel combination of $1/h_{oe}$ and R_C, but R_C is *external* to the transistor and the output resistance is the resistance seen looking back into

the transistor when the input is zero. *Careful here!* Remember for small signal operation $I_b = 0$ really means I_b held constant at its d.c. bias value. So when $I_b = 0$, $h_{fe} \times I_b = 0$

$$\therefore R_{out} \simeq \frac{1}{h_{oe}}.$$

Voltage gain A_v

$$A_v = \frac{V_{out}}{V_{in}} \quad Now: V_{in} = V_{be}$$

$V_{out} = -V_{ce}$ (because it is an inverting amplifier)

$$A_v = \frac{-V_{ce}}{V_{be}} \quad V_{be} = I_b h_{ie} \quad -V_{ce} = I_c R_c$$

$$I_c = h_{fe} I_b \ and \ I_b = \frac{V_{be}}{h_{ie}} \ \therefore I_c = h_{fe} \times \frac{V_{be}}{h_{ie}}$$

$$\therefore V_{ce} = \frac{-h_{fe} V_{be}}{h_{ie}} \times R_c$$

$$so \quad A_v = \left[\frac{-h_{fe} V_{be}}{h_{ie}} \times \frac{R_c}{V_{be}} \right] = \frac{-h_{fe} R_c}{h_{ie}}$$

$$A_v \simeq \frac{-h_{fe} R_c}{h_{ie}}$$

A Further Word About BJT Operation

It has been stated that the BJT is a current operated device in which the input current controls the output current

$$h_{fe} = \frac{\delta I_C}{\delta I_B} = \frac{I_c}{I_b}$$

However I_B causes a change in the input voltage (V_{BE}) and the reality is that a change in the input voltage (V_{BE}) brings about a change in output current (I_C). By considering that δV_{BE} causes δI_C we can determine the mutual conductance gm of the BJT. Thus:

$$gm = \frac{\delta I_C}{\delta V_{BE}} = \frac{I_c}{V_{be}} \ siemens \ (S)$$

Now since $h_{fe} = \dfrac{I_c}{I_b}$ and $h_{ie} = \dfrac{V_{be}}{I_b}$

We can write

$$\dfrac{h_{fe}}{h_{ie}} = \dfrac{I_c/I_b}{V_{be}/I_b} = \dfrac{I_c}{V_{be}} \therefore \dfrac{h_{fe}}{h_{ie}} = gm$$

So the voltage gain A_v of the common emitter amp. can be expressed as

$$A_v \simeq -gmR_C$$

showing that as with the FET the gain A_v is a function of the load resistor R_C — jolly convenient!

Self Assessment 3

---

1   A BC.177 is used in common emitter mode with a 4k7 collector resistor. If

   $h_{ie} = 1\,k$ $h_{fe} = 125$ $h_{oe} = 20\,\mu S$

   determine a) R_{in}
   b) R_{out}
   c) A_i
   d) gm
   e) A_v

2   A JFET is used in common source mode with a 8k2 load resistor. If

   $gm = 30\,mS$ and $rds = 40\,k\Omega$

   determine A_v

The Testing of Amplifiers

A number of simple tests can be performed on an amplifier in order to determine the following specifications:

- The quiescent current
- The gain
- The signal distortion level ⎫ All amplifiers
- The bandwidth
- The input and output resistance ⎭
- The maximum power output ⎫ Power
- Output signal distortion ⎭ amplifiers

Quiescent Current

This is the current that the circuit draws from the d.c. power supply when *no* input signal is applied.

Method

1   Connect the d.c. power supply to the amplifier circuit but connect an ammeter in the positive supply lead.
2   *Switch on*, measure and record the current drawn from the supply.

The Gain (Fig. 2.26)

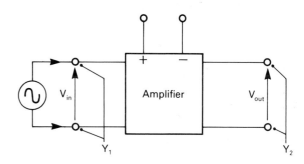

Fig. 2.26   Circuit to determine the gain

This refers to the voltage gain A_v which you already know to be:

$$A_v = \dfrac{V_{out}}{V_{in}} \therefore A_v = 20\,Log\left(\dfrac{V_{out}}{V_{in}}\right) dB$$

Method
1   Connect the amplifier to a power supply and adjust for the correct supply voltage.
2   Connect a signal generator to the input and adjust it to give a sine wave output at 1 kHz of about 20 mV peak-to-peak. This is V_{in}

THOUGHT _____

Why 1 kHz? This is a convenient frequency because it is not low enough to be influenced by any coupling capacitors, it is easy to get a clear stable trace on the CRO, and for audio amplifiers it is a standard test frequency.

3 Monitor V_{in} and V_{out} with the CRO (check V_{out} is undistorted, if it is, reduce V_{in}).
4 Measure and record the peak-to-peak values from the CRO screen and calculate A_v.

The signal distortion level

Method

1 Leave the arrangement as it was set up for the measurement of gain (Fig. 2.26).
2 While observing V_{out} on the screen, increase V_{in} (signal generator output) until V_{out} just starts to distort, i.e. clips or appears non-sinusoidal. Reduce V_{in} until V_{out} is just normal and measure V_{in}.
3 The value of V_{in} is now the maximum input signal that can be applied before distortion occurs.

The Bandwidth

This is the range of frequencies over which an amplifier can be used. It is further specified as the range of frequencies over which the gain falls by no more than 3 dB from its mid-band value. This can be a slightly tedious test to perform, but it is one of the most crucial, and if done correctly can be accomplished quickly. The test involves making a series of gain measurements from a very low frequency, say 10 Hz, to a very high frequency, say 1 MHz, and then plotting the results as a frequency/gain response curve (Fig. 2.27).

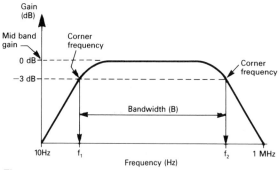

Fig. 2.27   Graph of gain against frequency

From this curve the upper and lower 'corner' frequencies f_1 and f_2 can be found and the bandwidth (B) established.
    Using $B = f_2 - f_1$,

if $f_1 = 25$ Hz and $f_2 = 980$ kHz
then $B = 980$ kHz $- 25$ Hz $\simeq 980$ kHz.

Method

1 Set up the equipment as it was for measuring the gain (Fig. 2.26).
2 Set V_{in} to a value well below the maximum level, e.g. if $V_{in}(max) = 50$ mV before distortion occurs, set it to 20 mV (say).
3 Prepare a table thus:

Frequency (Hz)	V_{in} (mV)	V_{out} (mV)	$A_v = 20 \, Log \left(\dfrac{V_{out}}{V_{in}} \right)$ dB
10	20	24	1.2
30	20	30	1.5

4 You are now going to make the series of measurements that will enable the graph to be drawn.

THOUGHT

Over what frequency steps are the measurements made?
If the amplifier has a flat response then it is only necessary to make a large number of measurements when the gain is changing (Fig. 2.28).

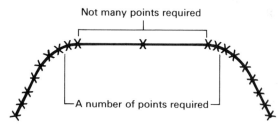

Fig. 2.28   Illustration of method for plotting a gain-frequency curve

So it is a good idea to sweep or scan the entire range from 10 Hz–1 MHz with the generator before recording any results just to see how the gain changes. Then from this broad overview you can decide where over the range to make the most measurements.

5 Make your measurements, calculate A_v and plot the curve.

Note   It is, I think, a good idea to plot the graph as you go. In this way any 'rogue' measurements will be immediately apparent and, since the equipment is connected, checking and repeating is simple. Otherwise you will be in the position of plotting your results from a table, and any errors will mean considerable work if they are to be corrected!

Plotting the curve

You will need log/lin graph paper for this, with the logarithmic scale along the horizontal axis and the linear scale vertical. This type of paper is available in cycles, e.g. 3 cycle log/lin paper (Fig. 2.29).

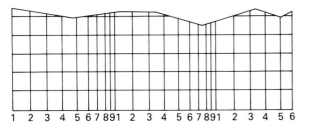

Fig. 2.29   Three cycle log/linear graph paper

The reason for this cramping of the frequency scale is to enable a large frequency range to be accommodated on a reasonably sized sheet of paper, and gives a meaningful picture of what is happening. *Notice the lowest figure is 1 and the highest 9*. This means you decide on the actual values, e.g. 1 could be 1 Hz then 9 would be 9 Hz the next 1 would be 10 Hz, 2 would be 20 Hz etc. So if you are plotting a response over the range of 10 Hz – 1.5 MHz you will need 6 cycle log/lin paper – marked up as shown below.

The gain once calculated is plotted up the side normally. When your points have been transferred they can be joined to form the curve, the mid-band value can then be established, let's say it is 50 dB. The bandwidth (B) is enclosed by the upper and lower frequency points where the gain is 3 dB down on 50 dB (50 dB = mid-band gain).

f_1 and $f_2 = 50 - 3 = 47$ dB

B can now be firmly established.

Input and Output Resistance

Perhaps the easiest method of estimating these is to use Ohm's Law and a variable resistance, preferably a decade resistance box in conjunction with the usual laboratory signal generator and oscilloscope. You will do well to remember that the accuracy of the result reflects the care and vigilance with which the testing has been carried out.

Input Resistance (Fig. 2.30)

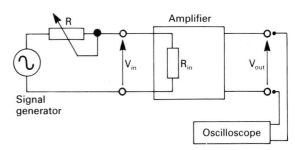

Fig. 2.30   Circuit to determine input resistance

1   Set R to zero ($0\,\Omega$ = short circuit).
2   Set the signal generator to 1 kHz sine wave and adjust to give a suitable V_{in}, i.e. well below the maximum permissible level before distortion.
3   Measure V_{out}.
4   Increase the setting of R until V_{out} falls to exactly half its original value.
5   The value of R is now approximately the input resistance of the amplifier.

Note   This is because R is now the same value as R_{in}, i.e. they are matched. Consequently half the input voltage is dropped across R and half across R_{in} $\therefore$ V_{out} will be half its original value.

Note   It is worth remembering that the input resistance of a common emitter amplifier will be approximately the quoted h_{ie} of the transistor. For a common source amplifier R_{in} will be approximately that of the gate resistor R_G.

10 Hz	90 Hz	100 Hz	900 Hz	1 kHz	9 kHz	10 kHz	90 kHz	100 kHz	900 kHz	1 MHz	9 MHz
1 cycle		2 cycles		3 cycles		4 cycles		5 cycles		6 cycles	

Output Resistance (Fig. 2.31)

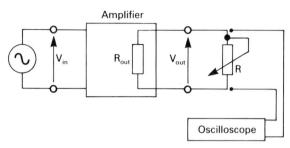

Fig. 2.31   Circuit to determine output resistance

1   Disconnect R and monitor V_{out} with the CRO.
2   Connect the signal generator and adjust to give a suitable value of V_{out} at 1 kHz.
3   Connect R and adjust until V_{out} is exactly half the value it was.
4   The value of R now represents the output resistance of the amplifier.

Note   This is a particularly difficult test to perform because of the comparatively low resistance of the amplifier output. If any distortion occurs in the output signal repeat the test with a lower value of V_{in}.

Power Amplifiers

Maximum output power

Power amplifiers deliver their maximum output power into a specified load (typically 8 Ω for an audio amplifier). This can be measured in the following way:

1   Set the amplifier volume control (if fitted) to maximum.
2   Monitor the voltage (V_{out}) across the load resistors.
3   A 1 kHz sine wave should be used as the input signal and its amplitude increased to give maximum V_{out} *without* distortion.

Thus calculation of maximum power can be achieved using Ohm's Law

$$P = \frac{(V_{out})^2}{R_{out}}$$ as shown in Fig. 2.32.

In practice R_{out} is usually a loudspeaker but most tests on audio power amplifiers are performed using a *dummy* load which is a resistor of the

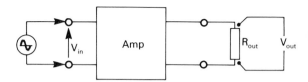

Fig. 2.32   Circuit to determine output power

same value as the speaker *and* capable of dissipating the required amount of power, i.e. if the amplifier has a maximum power output of 60 W into a speaker of 8 Ω impedance, the dummy load must be an 8 Ω resistor with a power rating of at least 60 W. Incidentally, the speaker itself could be used but at full power it is quite likely that the test engineer or his colleagues will go deaf or mad if the test takes some time.

When performing this test it is quite likely that a CRO will be used for measuring V_{out} because it allows a visual check for distortion to be made at full power. Remember however that output power is quoted using rms values; therefore you must convert the peak-to-peak output signal to V_{out} (rms).

$$V_{out}(rms) = \frac{V_{out}\ peak\text{-}to\text{-}peak}{2} \times 0.707$$

$$P_{out}(max) = \frac{(V_{out}(max)rms)^2}{R_{load}}$$

Signal Distortion

This is often called the *Total Harmonic Distortion* (THD) and is quoted as a percentage of the output signal, e.g. the integrated circuit power amplifier TDA 2004 has a THD of 10% at 1 kHz, this indicates the amount of distortion present in the output signal.

Harmonic distortion is very difficult to measure without specialized laboratory equipment like a *distortion factor meter*, a *harmonic analyser* or a *spectrum analyser*. It is therefore a quantity that is beyond the scope of this book and will alas go unmeasured.

The Tuned Circuit Amplifier

It is sometimes required to amplify only a very narrow band of frequencies centred on one particular frequency. Such amplifiers have the frequency response curve as shown in Fig. 2.33.

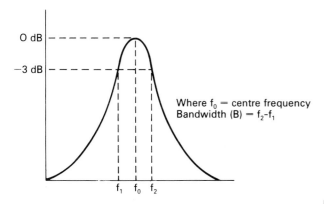

Fig. 2.33 Tuned amplifier frequency response

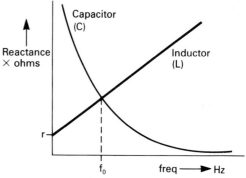

Fig. 2.35 Inductive and capacitive reactance

The frequencies either side of the selected band receive minimum amplification while the centre frequency (f_0) is amplified the most. Amplifiers of this type are employed wherever selectivity is required and often form the basis of radio tuning stages and band-pass filters.

From previous work we know that the voltage gain A_v of a common source and common emitter amplifier is determined by the load resistor since

$A_v \simeq -gm \times R_D$ for the common source amplifier
and
$A_v \simeq -gm \times R_C$ for the common emitter amplifier.

Therefore to achieve the desired frequency response the load resistor must be replaced by a circuit with a resistance that is frequency dependent, giving maximum resistance at f_0. A circuit that has just this characteristic is the parallel tuned circuit.

The parallel tuned circuit

This consists of an inductor (L) and capacitor (C) connected as shown in Fig. 2.34.

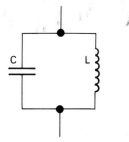

Fig. 2.34 Parallel tuned circuit

Both the components in this circuit exhibit the characteristic known as reactance (X ohms) which varies with frequency as shown in Fig. 2.35.

Note XL is offset by the resistance of the coil (r).

You can see that inductive reactance (XL) increases with frequency while capacitive reactance (XC) decreases with an increase in frequency. These reactance values can be calculated using

$$XL = \omega L \text{ ohms and } XC = \frac{1}{\omega C} \text{ ohms}$$

where $\omega = 2\pi f$

These reactive components together with the coil resistance combine to give the circuit impedance (Z ohms) which is a maximum at f_0. At frequencies below f_0 the low reactance of the inductor (L) *reduces* the circuit impedance while at frequencies above f_0 the low reactance of the capacitor (C) also *reduces* the impedance of the circuit.

The frequency (f_0) at which the impedance (Z) is maximum is known as the *resonant frequency* and can be determined in the following ways.

The Approximate Method

Consider the circuit shown in Fig. 2.36.

Theoretically the current in the capacitor (I_C) leads the applied voltage (V) by $90°$ while the current in the inductor (I_L) lags the applied

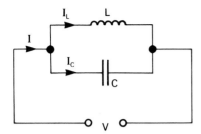

Fig. 2.36   Current in the tuned circuit

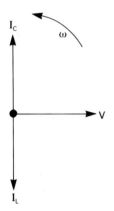

Fig. 2.37   Phasor diagramm of the ideal circuit

voltage (V) by 90° — providing the resistance of the inductor is zero!

This implies that at resonance (f_o), $I_C = I_L$ and that these currents will be equal and opposite and thus cancel. So at only one frequency will the applied current (I) be in phase with the applied voltage (V).

This resonant frequency can be found in the following way: at resonance, $XL = XC$.

$$XL = 2\pi f_o L \quad XC = \frac{1}{2\pi f_o C}$$

$$\therefore 2\pi f_o L = \frac{1}{2\pi f_o C}$$

cross multiply to give $1 = 2\pi f_o L \times 2\pi f_o C$
multiply out $1 = 4\pi^2 f_o^2 LC$
make f_o^2 subject of formula:

$$f_o^2 = \frac{1}{4\pi^2 LC}$$

$$\therefore f_o = \frac{1}{2\pi \sqrt{LC}} \text{ Hz}$$

CAUTION! This is actually the formula for determining the resonant frequency of a series *LC* circuit and the one we are using is a parallel arrangement. However it serves as a useful approximation for the resonant frequency of a parallel tuned circuit and this is why you will often find it being widely used.

THOUGHT

Why is it not quite right? — The impedance of a parallel *LC* circuit must take into consideration the resistance of the coil itself, this in turn will have an effect on the resonant frequency (f_o). In the series circuit however the coil resistance (r) will not effect f_o.

Accurate Calculation of f_o

The inductor coil possesses resistance (r). See Fig. 2.38.

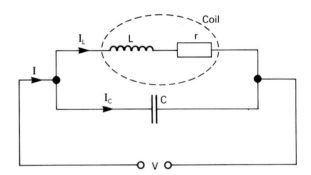

Fig. 2.38   Actual tuned circuit

This will cause I_L to lag the voltage (V) by an angle less than 90° and so the value of the coil resistance will affect the frequency (f_o) at which the applied voltage (V) and current (I) are in phase. This is shown in Figs 2.39 (a), (b) and (c).

The actual resonant frequency can be accurately determined using:

$$f_o = \frac{1}{2\pi} \sqrt{\frac{1}{LC} - \frac{r^2}{L^2}} \text{ Hz}$$

An example here will serve as a good way of comparing the two methods.

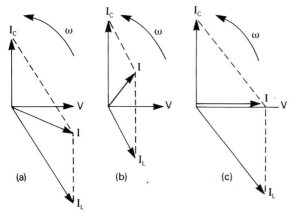

Fig. 2.39 (a) Freq. below f_o (I lags V); (b) Freq. above f_o (I leads V); (c) Freq. $= f_o$ (I in phase with V)

Example

A 10 mH inductor with a coil resistance (r) of 50 Ω is connected in parallel with a 0.1 μF capacitor. Calculate the resonant frequency using a) The approximate method and b) The equation for parallel resonance.

a) $\quad f_o = \dfrac{1}{2\pi \sqrt{LC}}$

$\quad = \dfrac{1}{6.28 \sqrt{10 \times 10^{-3} \times 0.1 \times 10^{-6}}}$

$\quad = 5035\ \text{Hz}$

b) $\quad f_o = \dfrac{1}{2\pi} \sqrt{\dfrac{1}{LC} - \dfrac{r^2}{L^2}}$

$\quad = \dfrac{1}{6.28} \sqrt{\dfrac{1}{(10 \times 10^{-3} \times 0.1 \times 10^{-6})}}$

$\qquad - \dfrac{50^2}{(10 \times 10^{-3})^2}$

$\quad = 4972\ \text{Hz}$

from this you can see that there is a difference and the approximation will only be useful when

(a) the coil resistance is very low. (When compared to LC).

(b) the coil resistance is unknown and cannot be measured.

(c) a rough guide to f_o is required, e.g. when designing a circuit.

THOUGHT

So the greater the coil resistance the more inaccurate the approximation will be? — Yes.

The parallel tuned circuit can be used in place of the load resistor in the common emitter and common source amplifier as shown in Figs 2.40 and 2.41.

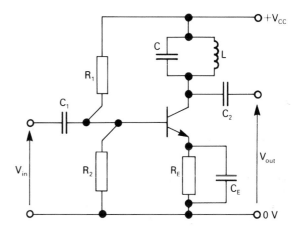

Fig. 2.40 Tuned collector amplifier

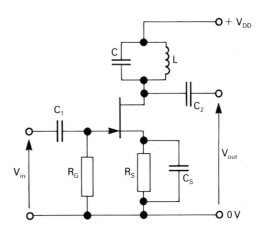

Fig. 2.41 Tuned drain amplifier

For these amplifiers the frequency at which the voltage gain (A_v) will be maximum can be estimated using:

$$f_o = \frac{1}{2\pi\sqrt{LC}}\,\text{Hz}$$

Note In the absence of any information about the coil resistance the approximation has to be used.

Calculation of A_v

The voltage gain will be determined by the impedance (Z) of the tuned circuit at resonance and thus $A_v = -gm \times Z$. It has already been stated that the actual impedance (Z) is a combination of XC, XL and r of the parallel tuned circuit. Now at resonance, XC and XL are equal and opposite so the impedance will be determined by the coil resistance r. At resonance the actual impedance is purely resistive and is called the dynamic impedance (Z_d) given by

$$Z_d = \frac{L}{Cr}$$

where L = coil inductance in henrys
C = capacitance in farads
r = d.c. coil resistance in ohms

Example
A transistor with $gm = 200\,\text{mS}$ has a tuned circuit load where $L = 1\,\text{mH}$, $C = 1\,\mu\text{F}$ and $r = 4\,\Omega$. Calculate

(a) the impedance of the load at resonance.
(b) the voltage gain A_v at resonance.

(a) $Z_d = \dfrac{L}{Cr} = \dfrac{1 \times 10^{-3}}{1 \times 10^{-6} \times 4} = 250\,\Omega$

(b) $A_v = -gm \times Z = 200 \times 10^{-3} \times 250 = 50$

$A_v = 50$ (in dB, $A_v = 20\,\text{Log}\,50 = 34\,\text{dB}$).

Bandwidth (*B*)

The characteristic frequency response curve for a tuned amplifier is a bell-shaped curve with the

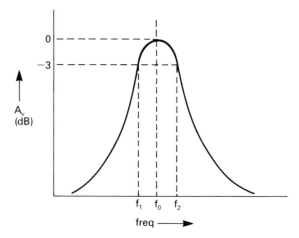

Fig. 2.42 Bandwidth of a tuned amplifier

bandwidth determined by the upper and lower -3 dB frequencies as shown in Fig. 2.42.

You will see from this that the actual bandwidth (B) is determined by the sharpness of the curve, i.e. the sharper the curve the narrower will be the bandwidth. Obviously when a very narrow bandwidth amplifier is required the curve must be very sharp. This characteristic is determined by the Q factor of the coil used in the tuned circuit.

Q Factor

This refers to the magnification quality of the coil and is given by

$$Q = \frac{2\pi f_o L}{r} \quad \text{(where } r \text{ is the resistance of the coil).}$$

From this you can see that the larger the value of coil resistance the lower will be the Q value. The higher the Q factor the sharper will be the response curve. See Fig. 2.43.

If the Q factor of the circuit is known, the bandwidth can be estimated using

$$B = f_2 - f_1 = \frac{f_o}{Q}.$$

f_1 and f_2 are the frequencies at which the voltage

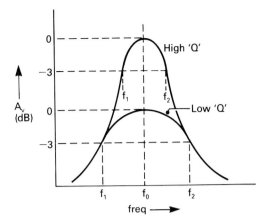

Fig. 2.43   *Q* factor and bandwidth

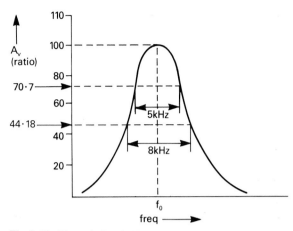

Fig. 2.44   The gain-bandwidth product

gain (A_v) falls to 0.707 of its value at f_o (i.e. 3 dB down from A_v at f_o). For our existing example:

$$f_o = 4.972 \text{ kHz} \quad L = 1 \text{ mH} \quad r = 4 \, \Omega$$

$$Q = \frac{2\pi f_o L}{r}$$

$$= \frac{6.28 \times 4.972 \times 10^3 \times 1 \times 10^{-3}}{4} = 7.8$$

$$B = f_2 - f_1 = \frac{f_o}{Q} = \frac{4.972 \times 10^3}{7.8} = 637.4 \text{ Hz}$$

This clearly indicates how in this circuit the *Q* factor determines the bandwidth. If the bandwidth is too narrow it can be widened by *reducing* the *Q* factor: this phenomenon is highlighted by a concept called the gain-bandwidth product.

The gain-bandwidth product

Consider the frequency response indicated in Fig. 2.44.

If the bandwidth (B) = 5 kHz and the numerical value of gain (A_v) = 70.7, the gain-bandwidth product GB = Gain (A_v) $\times B$

$$\begin{aligned} &= 70.7 \times 5 \times 10^3 \\ &= 353.5 \text{ kHz} \end{aligned}$$

This is a constant!

$$\therefore \text{ if } GB = A_v \times B$$

$$\text{then } B = \frac{GB}{A_v} \text{ and } A_v = \frac{GB}{B}$$

THOUGHT

So to widen the bandwidth the gain is reduced BUT the gain-bandwidth product (GB) remains the same? Yes — but note that once the GB is known, the required gain or bandwidth can be estimated.

So if a bandwidth of 8 kHz is required the *gain A_v* should be

$$\frac{353.5 \times 10^3}{8 \times 10^3} = 44.18.$$

The simplest way to reduce the gain and thus increase the bandwidth is to add a resistor (R) in series with the inductor (L) of the tuned circuit to lower or *dampen* its *Q* factor, as shown in Fig. 2.45.

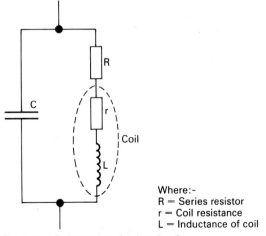

Where:-
R = Series resistor
r = Coil resistance
L = Inductance of coil

Fig. 2.45   Series resistor for damping *Q*

PRACTICAL INVESTIGATION **2**

The Tuned Collector Amplifier

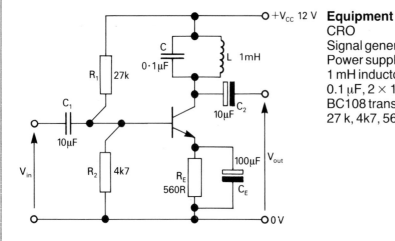

Equipment
CRO
Signal generator
Power supply
1 mH inductor
0.1 µF, 2 × 10 µF, 100 µF capacitors
BC108 transistor
27 k, 4k7, 560R, 22R resistors

Method
1 Build the circuit as shown.
2 Calculate the resonant frequency of the tuned circuit.
3 Monitor V_{in} and V_{out} with the CRO.
4 Connect the signal generator and adjust to give a sine wave V_{in} of 50 mV peak-to-peak at the calculated resonant frequency.
5 Observe the output and *swing* the signal generator frequency back and forth to determine the exact frequency at which gain is maximum. This is the actual resonant frequency — record this.
6 Calculate the voltage gain at resonance in decibels.
7 Select a suitable frequency range and plot the gain/frequency response of the amplifier.
8 Place a 22R resistor in series with the inductor and repeat steps 5, 6 and 7.

Results
1 Is the actual resonant frequency the same as the calculated value — justify your findings.
2 From your graphs determine the bandwidth of each circuit and find the gain-bandwidth product.
3 Make a general observation of the relationship between gain and bandwidth.
4 Using information from your graphs calculate the Q factor of the circuit in both cases.
5 Comment on the relationship between Q factor and bandwidth.

DESIGN ASSIGNMENT 1

A BJT is to be used as a tuned amplifier. The transistor itself has the following specifications:

$h_{fe} = 180\,(\text{min})$ $h_{ie} = 1.5\,\text{k}$.

The finished amplifier is required to be tuned to 12 kHz and have a 600 Hz bandwidth. A 1 mH inductor is available that has a coil resistance of 2 Ω.

Design a circuit and show how the appropriate component values have been selected.

The tuned amplifier practical investigation will enable you to confirm these theories.

Procedure

Starting point
As always start with the known and work towards the unknown

1 Start by considering the tuned circuit

$L = 1\,\text{mH}$ $r = 2\,\Omega$ $f_o = 12\,\text{kHz}$ $C = ?$

Calculate the value of C required for resonance of 12 kHz:

$$\text{Since } f_o \simeq \frac{1}{2\pi \sqrt{LC}} \qquad C = \frac{1}{(2\pi f_o)^2 L}$$

Note   Although we know the coil resistance we use the approximation for simplicity and because the resistance is very low.

$$C = \frac{1}{(6.28 \times 12 \times 10^3) \times 1 \times 10^{-3}} = 176\,\text{nF}$$

2 Determine the dynamic impedance (Z_d).

$$(Z_d) = \frac{L}{Cr} = \frac{1 \times 10^{-3}}{176 \times 10^{-9} \times 2} = 2.84\,\text{k}\Omega$$

3 Calculate A_v

$$A_v = -gmZ$$

$$\text{and } gm = \frac{h_{fe}}{h_{ie}} = \frac{180}{1.5 \times 10^3} = -120\,\text{mS}$$

(remember the minus indicates phase inversion)

$$A_v = -120\,\text{mS} \times 2.84\,\text{k}\Omega$$
$$= -120 \times 10^{-3} \times 2.84 \times 10^3$$
$$= -340.8.$$

$$A_v = -341.$$

4 Calculate Q

$$Q = \frac{2\pi f_o L}{r} = \frac{6.28 \times 12 \times 10^3 \times 1 \times 10^{-3}}{2}$$

$$= 37.7.$$

$$Q = 37.7.$$

5 Determine the bandwidth (B)

$$B = f_2 - f_1 = \frac{f_o}{Q} = \frac{12 \times 10^3}{37.7} = 318\,\text{Hz}$$

$$B = 318\,Hz$$

The specifications call for a bandwidth of 600 Hz so it is now time to consider how to change the bandwidth. First state all the relevant information.

Specified bandwidth = 600 Hz
Calculated bandwidth = 318 Hz
Maximum gain $(A_v \text{ at } f_o) = 341$.

6 Determine the actual gain A_v that produces the bandwidth of 318 Hz

When $B = 318$ $A_v = 341$

7 Calculate the gain-bandwidth product (*GB*)

$$GB = gain\ (A_v) \times bandwidth$$
$$= 341 \times 318 = 108.4$$

8 Calculate the gain required to give $B = 600\ \text{Hz}$

$$GB = A_v \times B \qquad A_v = \frac{GB}{B}$$

$$= \frac{108.4 \times 10^3}{600} = 180.7$$

$$A_v \simeq 181$$

So to produce the required bandwidth it is necessary to reduce the voltage gain of the circuit from 341 to 181. This may seem a difficult thing to achieve at this point since we are contemplating the theoretical values of a design problem. A solution is available when you realize that the *Q* factor determines the bandwidth and consequently must affect the gain.

9 Determine the *Q* factor required to give $B = 600\ \text{Hz}$

$$Q = \frac{f_o}{B} \qquad f_o = 12\ \text{kHz}$$
$$B = 600\ \text{Hz}$$

$$\therefore \qquad Q = \frac{12 \times 10^3}{600} = 20$$

To widen the bandwidth to 600 Hz the *Q* must be reduced to 20. This is achieved by adding a resistor in series with the coil in the tuned circuit.

10 Calculate *R* to be added in series with the coil.

$$Q = \frac{2\pi f_o L}{r + R} \qquad r + R = \frac{2\pi f_o L}{Q}$$

$$r + R = \frac{6.28 \times 12 \times 10^3 \times 1 \times 10^{-3}}{20}$$

$$= 3.77\ \Omega$$

since $r = 2\ \Omega$ the series resistor $R = 1.77\ \Omega$ (nearest preferred value 1.8 Ω).

The design procedure has deliberately included a *blind alley* to show that the apparent solution to one part of the problem may make it impossible to move to the next stage. Find an alternative solution however and things may fall into place very easily. Please remember all the values and the methods used to find them are the result of theoretical concepts. The next step is to build and test the designed circuit and then make such practical adjustments as may be necessary to meet the desired specifications.

More About Tuned Amplifiers

Variable tuning

It is the *LC* network that determines the amplifier's resonant frequency f_o, consequently if either *L* or *C* could be adjusted it would be possible to vary f_o as shown in Fig. 2.46(a).

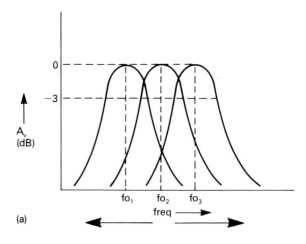

(a)

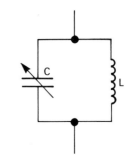

(b)

Fig. 2.46   (a) Variable tuning; (b) Variable tuned circuit

It is this principle that makes a tuned amplifier particularly suitable for a radio frequency application, where it enables one frequency to be selected to the exclusion of all others. Adjustability is easily achieved with a variable capacitor as shown in Fig. 2.46(b).

Biasing a Tuned Amplifier

It was mentioned in the previous section on amplifier biasing that Class C bias can be employed for tuned amplifiers. If this is so the distortion produced by such biasing must somehow be overcome. To understand how, let's take a closer look at the tuned circuit of Fig. 2.47.

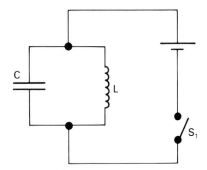

Fig. 2.47   CL circuit

If S_1 is closed current flows in L and a magnetic field is set up. If S_1 is then opened current flow stops, the magnetic field collapses setting up a back EMF that charges C. The back EMF will die out and C will then discharge, causing current to flow in L. This in turn will set up a magnetic field which will collapse when C is discharged, the back EMF will charge C. The result will be the damped oscillation waveform at the frequency f_o as shown in Fig. 2.48. *Notice* that although the waveform diminishes it is a constant frequency sine wave. Now if energy could be supplied to the tuned circuit at the same rate as it is lost, the sine wave will have a constant amplitude — this is exactly what the Class C amplifier does.

The *blips* of current I_c stimulate the tuned circuit and a sine wave output is produced. When the input signal is the same as the resonant frequency (f_o) the output signal will be a maximum amplitude sine wave even though the current

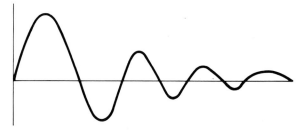

Fig. 2.48   Damped oscillation

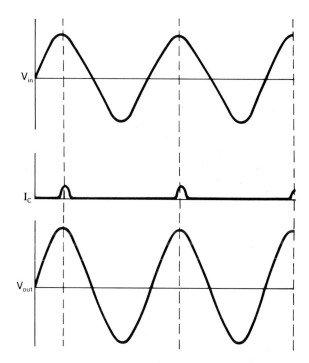

Fig. 2.49   Class C tuned amplifier waveforms

flowing in the collector is not a sine wave but simply a series of current pulses occurring at f_o.

Amplifier Review

1   There are small signal and large signal (power) amplifiers.
2   A transistor can be biased to operate in Class A, AB, B or C.
3   Class A bias means that current flows in the output for a full 360° of input signal, giving low distortion but poor efficiency.

4   Class B bias allows current to flow in the output for only 180° of input signal, giving better efficiency but distortion due to the non-linear part of the transistor input characteristic.
5   Class AB is a compromise between Classes A and B. Current flows in the output for more than 180° but less than 360° giving lower efficiency than Class B but a less distorted output signal.
6   Class C bias is the most efficient of all because current flows in the output for less than 180° of input signal but the distortion produced is very great.
7   A d.c. load line enables the quiescent (Q) point to be chosen.
8   The a.c. load line shows how the device operates under signal conditions.
9   An amplifier's operation can be estimated using *small signal* equivalent circuits.
10 An h parameter equivalent circuit can be used to model the performance of a Bipolar Junction Transistor (BJT).
11 Capacitors and d.c. power supplies are considered to be short circuit under signal conditions.
12 For a common emitter amplifier $R_{in} \simeq h_{ie}$ $A_i \simeq h_{fe}$

$$A_v \simeq \frac{-h_{fe}R_C}{h_{ie}} \qquad R_{out} \simeq \frac{1}{h_{oe}}$$

13 Although a current operated device; the mutual conductance (gm) of the BJT can be found using

$$gm = \frac{h_{fe}}{h_{ie}}$$

14 The Field Effect Transistor (FET) is a voltage operated device with the parameters: rds = drain source resistance, gm = mutual conductance.
15 For a common source amplifier:
$R_{in} \simeq R_G$
$R_{out} = rds \| R_D$
($\|$ means in parallel with)
$A_v \simeq gmR_D$ (where R_D = Drain resistor)
16 The voltage gain A_v of a common emitter amplifier can be found using $A_v \simeq gmR_C$ (where R_C = Collector resistor).

17 A tuned circuit amplifier has maximum gain at only one frequency (f_o).
18 A parallel tuned circuit has maximum impedance at resonance, giving $A_v \simeq gmZ$ if it takes the place of the collector or drain resistor in a common emitter and common source amplifier.
19 The dynamic resistance of a parallel tuned circuit is given by

$$Z_d = \frac{L}{Cr}$$

20 The Q factor of the coil in the tuned circuit determines the bandwidth of the amplifier. High Q = narrow bandwidth.
21 The Q factor is determined by the coil resistance (r)

$$Q = \frac{2\pi f_o L}{r}$$

22 Bandwidth, $B = \frac{f_o}{Q}$.

23 The *gain × bandwidth* (GB) product of a tuned amplifier is a constant. GB = numerical value of A_v × *bandwidth*.
24 Adding additional resistance in series with the coil of a tuned amplifier dampens Q thus reducing the gain and widening the bandwidth.
25 The frequency of a tuned amplifier can be altered by varying either the capacitor or inductor of the tuned circuit.
26 Class C biasing can be used for a tuned amplifier, with the tuned circuit removing the characteristic distortions.

Self Assessment Answers

Self Assessment 2

d.c. load line

$V_{DS} \simeq V_{DD}$ = 10 V = point A

$$I_C = \frac{V_{DD}}{R_D + R_S} = \frac{10}{800 + 200} = \frac{10}{1000}$$

= 10 mA = point B

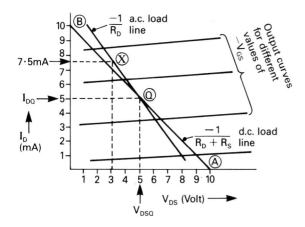

a.c. load line

$$\text{slope} = -\frac{1}{R_D} = \frac{1}{800} = 1.25 \text{ mA/V}$$

∴ if V_{DS} falls by 2 V, I_D increases by
2×1.25 mA = 2.5 mA
So $V_{DSQ} = 5$ V
If V_{DS} falls by 2.0 V $\quad V_{DS} = 5 - 2 = 3$ V
then
$I_{DQ} = 5$ mA this will increase by 2.5 mA
∴ $I_D = 7.5$ mA
The a.c. load line point x is where
$V_{DS} = 3$ V and $I_D = 7.5$ mA.

Self Assessment 3

1

(a) $R_{in} \simeq h_{ie} = 1 \text{ k}\Omega$

(b) $R_{out} \simeq \dfrac{1}{h_{oe}} = 50 \text{ k}\Omega$

(c) $A_i \simeq h_{fe} = 125$

(d) $gm = \dfrac{h_{fe}}{h_{ie}} = \dfrac{125}{1 \text{ k}} = 125 \text{ mA/V}$

(e) $A_v = -gmR_C = -587.5$

$\quad$ or $A_v = \dfrac{-h_{fe}R_C}{h_{ie}} = -587.5$

2 $A_v = -gmR_D$
$\quad = 30 \text{ mS} \times 8k2 = -246$

3

Power Amplifiers

The main task of a power amplifier is to boost its input signal to a level that is high enough to do physical work, for example operate some form of electro-mechanical transducer like a loudspeaker or motor, or produce electromagnetic radiation from a transmitting aerial. The amplifier must achieve this with due consideration given to the following:

1   At no time must the maximum specifications for any circuit component be exceeded.
2   Distortion must be kept to a minimum.
3   The efficiency of the circuit must be as high as possible.

There are a vast number of circuits possible when considering the design of power amplifiers but as always some 'favourites' emerge and become virtually standard circuits. This is usually because they embody the best compromise when all the factors have been considered. In this section we shall examine the popular audio power amplifier in order to develop a familiarity with the concepts behind power amplifiers.

To achieve the required amount of power output from any amplifier, maximum use must be made of the largest possible changes in voltage and current signals that are flowing in the amplifying devices. Hence the alternative name of *large signal amplifiers*.

Let us consider the single BJT used as a power amplifier (Fig. 3.1).

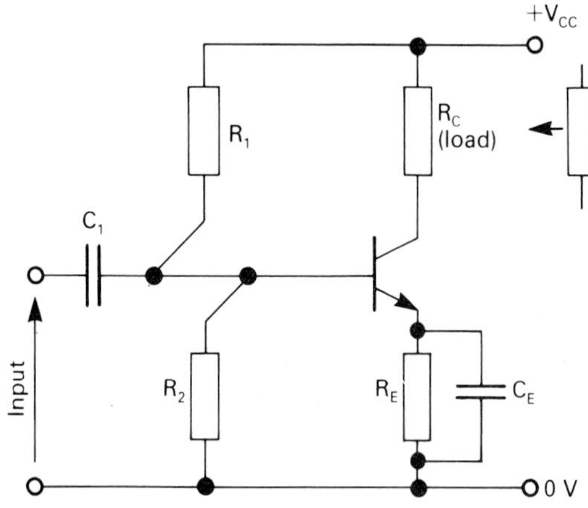

Since only one transistor is used this is called the single-ended amplifier.

THOUGHT

Where is the output taken from? The output current in the transistor is the collector current (I_c). Therefore in this circuit the load itself is connected in place of R_c.

Now if the load was a loudspeaker it could directly replace R_c as indicated but this would be a very poor arrangement for the following reason: For maximum current swings with minimum distortion the transistor would have to be biased in Class A. Therefore, a quiescent d.c. current (I_{CQ}) would always flow through the speaker coil, thus making it magnetically saturated by the d.c. current and unable to work properly. A slightly more successful arrangement would be to use a transformer where the primary winding replaces R_C and then d.c. flows in this rather than the speaker coil since this is now connected to the secondary winding. The transformer thus prevents the d.c. from passing through the loudspeaker. (See Fig. 3.2)

Fig. 3.1   Single transistor power amplifier

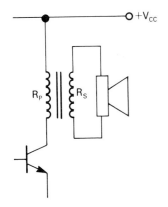

Fig. 3.2 Transformer coupling

Power Amplifier Load Lines

We have already considered the d.c. and a.c. load lines for small signal amplifiers; these load lines, however, will be slightly different for power amplifiers.

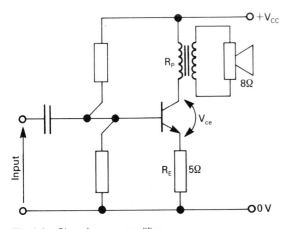

Fig. 3.3 Class A power amplifier

Studying the circuit of Fig. 3.3 there are subtle differences that separate this from its small signal companion. The load resistor (R_P) is the primary winding of a transformer but since it is a coil of wire its resistance will probably be quite a low value. The emitter resistor R_E is also a low value and has no de-coupling capacitor. *Why?* In this circuit R_E is only 5 ohms — imagine the huge

value capacitor required to bypass this low resistance at audio frequencies! For this reason it is common practice to leave the emitter resistor in power amplifiers un-decoupled.

The d.c. load line

Normally two points are required to draw a load line, these are determined using the equation

$$V_{CE} = V_{CC} - (R_E + R_P)I_C$$

but with this circuit the low resistances R_P and R_E are assumed to be negligible so the d.c. load on the transistor is zero!

This means the d.c. load line can be drawn vertically from the value $V_{CE} \simeq V_{CC}$ as shown in Fig. 3.4.

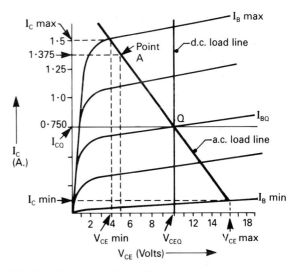

Fig. 3.4 Power amplifier load line

The Q point will be determined by R_1 and R_2 as usual to give Class A bias, thus ensuring that half the maximum collector current I_C flows under no signal conditions. Therefore, under quiescent conditions:

$$V_{CEQ} \simeq V_{CC} \quad \text{and} \quad I_{CQ} = \frac{I_C(\text{max})}{2}$$

The a.c. load line

The actual a.c. resistance of the transformer primary will be the resistance *looking in* to the primary winding given by

$$R_{in} = n^2 R_L$$

where n = turns ratio of the transformer
R_L = actual speaker or load resistance (which is reflected back to the primary)

The a.c. load line will have a slope of

$$-\frac{1}{R_{in}} = \frac{1}{n^2 R_L}$$

e.g. if n = 1:1 and R_L = 8 Ω

$$\text{slope} = -\frac{1}{8} = -0.125$$

This means −0.125 A/V indicating that for every 1 volt reduction in V_{CE}, I_C increases by 0.125 A. (The minus sign indicates that V_{CE} falls as I_C increases.)

To draw the a.c. load line only one other point is required since the Q point exists already, e.g. if V_{CEQ} = 10 V and I_{CQ} = 750 mA (Fig. 3.4).

When V_{CE} falls by 5.0 V, I_C will increase by 5×0.125 A = 0.625 making I_C = 1.375 A when V_{CE} = 5.0 V *POINT A*. The line can now be drawn using point A and the Q point.

It can now be seen that to use the maximum possible voltage and current swing the transistor must be driven between *cut-off* and *saturation*. Although the efficiency will be 50% the distortion will be severe, so in practice the swings are reduced, improving distortion but reducing the efficiency to about 25% as described in the section on amplifier biasing in Chapter 2.

THOUGHT

From the output characteristics and the load line it appears that the Q point can swing between 0 V and almost twice V_{CC}. *How can this be?* This effect is caused by the transformer primary winding. The collector current flowing in the winding induces an EMF and the induced EMF increases V_{CE} above V_{CC} as the collector current decreases.

Amplifier Efficiency

This refers to the conversion of energy from d.c. to a.c. The Greek letter 'eta' (η) is used to represent efficiency:

$$\eta = \frac{\text{a.c. output power}}{\text{d.c. input power}} \times 100\% = \frac{P_{ac}}{P_{dc}} \times 100\%$$

the a.c. output power is the power delivered to the load and the d.c. input power is the power supplied by the d.c. power supply. The reality is that not all the input d.c. power will be converted into usable a.c. output power. This is because some losses will inevitably occur.

$$P_{dc} = P_{ac} + P_L$$

where

P_{dc} = d.c. input power
P_{ac} = a.c. output power
P_L = power losses

The majority of power losses occur as collector power-dissipation losses, P_C.

$$\therefore \text{ if } P_L = P_C \quad P_{dc} = P_{ac} + P_C$$

and $P_C = V_{CE} \times I_C$

Example
Under operating conditions an amplifier draws 800 mA from its 10 V d.c. supply. If 6 W of audio output power is delivered to a loudspeaker. Calculate

(a)   the d.c. power (P_{dc}).
(b)   the collector power dissipation (P_C).
(c)   the efficiency of the amplifier (η).

(a)   d.c.   power   (P_{dc}) = d.c. current × d.c. voltage = 800 mA × 10 V = 8 W.
(b) $P_{dc} = P_{ac} + P_C$ ∴ $P_C = P_{dc} - P_{ac}$. P_{ac} = 6 W ∴ P_C = 8 − 6 = 2 W.

(c) $\eta = \dfrac{P_{ac}}{P_{dc}} \times 100 = \dfrac{6}{8} = 75\%$

From this example it becomes clear that the greater the losses that occur the more inefficient will be the power amplifier.

Power dissipation in the transistor

Most power is dissipated in the collector region of the transistor, at any point along the load line the power dissipation can be calculated using:

$$P_C = V_{CE} \times I_C$$

where

P_C = collector power in (watts)
V_{CE} = collector-emitter voltage (volts)
I_C = collector current (amps)

Maximum swings of current (I_C) and voltage (V_{CE}) must occur, but it is important to realize that at no time must the collector power (P_C) exceed the maximum rating for the chosen device (P_{tot} or P_C(max)).

To understand how the collector power changes during amplification study Fig. 3.5. This shows a transistor biased as a Class A power amplifier. The quiescent point has been selected and the load line drawn on the characteristic.

The quiescent values are: $I_{CQ} = 300$ mA and $V_{CEQ} = 6$ V. This gives a quiescent collector power $P_{CQ} = I_{CQ} \times V_{CEQ} = 300 \times 10^{-3} \times 6 = 1.8$ W.

(Remember this is wasted or lost power.)

When the input signal is applied, the current and voltage levels will change.

I_C varies from 100 mA to 500 mA.
V_{CE} varies from 2.5 V to 9.5 V.

When I_C is maximum at 500 mA, V_{CE} is minimum at 2.5 V.
When I_C is minimum at 100 mA, V_{CE} is maximum at 9.5 V.

Note These values indicate peak-to-peak swings! To determine the a.c. power developed in the collector, rms values must be used.

$$P_{ac} = I_C(rms) \times V_{CE}(rms)$$

$$I_C(rms) = \frac{I_C(max) - I_C(min)}{2} \times 0.707$$

$$= \frac{500 \text{ mA} - 100 \text{ mA}}{2} \times 0.707$$

$$= 282.8 \text{ mA}$$

$$I_C(rms) = 282.8 \text{ mA}$$

$$V_{CE}(rms) = \frac{V_{CE}(max) - V_{CE}(min)}{2} \times 0.707$$

$$= \frac{9.5 - 2.5}{2} \times 0.707 = 4.95 \text{ V}$$

$$V_{CE}(rms) = 4.95 \text{ V}$$

$$P_{ac} = 282.8 \times 10^{-3} \times 4.95 = 1.4 \text{ W}$$

$$P_{ac} = 1.4 \text{ W}$$

It may appear that since the a.c. collector power is 1.4 W a transistor with $P_{tot} = 1.5$ W will be suitable. *BUT* the transistor dissipates maximum power under quiescent conditions and then $P_{CQ} = 1.8$ W. So a device capable of handling a collector current greater than this must be used e.g. $P_{tot} = 2.0$ W.

To illustrate the variation in power a collector power dissipation curve (Fig. 3.5) can be drawn for the chosen device; this is achieved by using P_C(max) = $I_C \times V_{CE}$ to give the points to draw the curve.

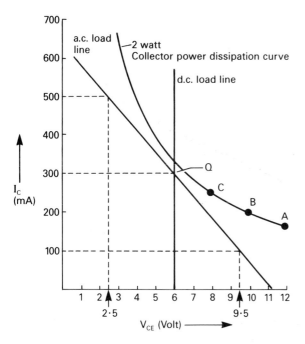

Fig. 3.5 Collector power curve

Example
Suppose a transistor with $P_{tot} = 2.0\,W$ is chosen.

$$P_C(max) = 2.0\,W$$

When $V_{CE} = 12\,V$

$$I_C = \frac{P_C}{V_{CE}} = \frac{2.0}{12} = 166.6\,mA \text{ (point A)}$$

When $V_{CE} = 10\,V$

$$I_C = \frac{2.0}{10} = 200\,mA \text{ (point B)}$$

When $V_{CE} = 8\,V$

$$I_C = \frac{2.0}{8} = 250\,mA \text{ (point C)}$$

etc. etc. until sufficient points have been plotted to enable a smooth curve to be drawn.

In this way a curve can be produced to show the maximum power dissipation of a chosen device. This indicates the *region of safety*. At all times during the operating cycle the current and voltage swings must be below this curve.

This is interesting because it serves to highlight the fact that with certain bias conditions the maximum collector power is dissipated under *no see how much*. Consider a single transistor with $P_{dc} = 3.6\,W$ and $P_{ac} = 1.4\,W$.

Under quiescent conditions only d.c. signals are present $\therefore P_{CQ} = P_{dc}$. While under operating conditions:

$$P_{dc} = 3.6\,W \qquad P_{ac} = 1.4\,W$$
$$P_{dc} = P_{ac} + P_C \text{ (where } P_C = \text{collector power losses)}$$
$$P_C = P_{dc} - P_{ac} = 3.6 - 1.4 = 2.2\,W.$$

So although the total collector power dissipation is 3.6 W, 2.2 W of this is wasted power — hardly very efficient!

Heat Sinks

The best designed amplifier with minimum loss will have some current flowing in the transistor when it is operating. This means that heat will be generated. If an amplifier is delivering power to a load, some power will be dissipated in the transistors themselves — if the devices are to remain stable they must be kept cool. For this reason all power amplifiers use power transistors capable of handling the current. To keep the internal temperature to a safe level a heat sink is used that effectively removes the excess heat. Modern heat sinks are made from aluminium with a matt black finish, they are bolted or clipped to the transistor and remove the heat by conduction and radiation.

THOUGHT

Why do most heat sinks have a matt black finish?
Heat is radiated more readily by a *black* body than one that is finished in any other colour.

Obviously the amount of cooling depends upon the surface area of the heat sink, so to create the maximum area in a small physical space the design usually incorporates fins, rather like those on a motor cycle engine. There are many different types of heat sinks available with a selection shown on page 152.

Choosing a heat sink

If you study the heat sink data on page 152 you will see that the specification quoted for each type is: Temperature (°C)/Watt. This is the thermal resistance of the device and indicates its ability to dissipate heat. The larger the heat sink the lower will be its thermal resistance and consequently the greater will be its ability to cool the power component that is mounted on it, e.g. compare the thermal resistance (R_{th}) of heat sink type 175-650 to that of 170-500.

175-650 $R_{th} = 13°C/W$
170-500 $R_{th} = 0.5°C/W$

With electrical conductivity a low electrical resistance indicates a good conductor. The same concept is used for thermal conductivity, i.e. a low thermal resistance for a heat sink indicates that it is a good conductor of heat and so will provide good cooling. As always cost will be involved and unfortunately the types with the lowest thermal resistance tend to be expensive because they are intricately shaped in order to provide maximum surface area for heat dissipation. The type of heat sink required will be determined by:

- The power dissipated by the electronic component.
- The maximum permissible working temperature.
- The ambient air temperature.

It is the actual junction temperature inside the transistor that is important. For this reason the metal case or tag is connected to the region of the semiconductor that will become the hottest (the collector/base region in a power transistor). You will remember that silicon devices can operate at quite high temperatures so a maximum functioning junction temperature (T_j) is usually assumed to be about 150°C.

So the *hot* silicon junction is connected to the transistor metal case or tag and this is fixed to the heat sink. Wherever a *joint* occurs a thermal resistance will exist and this must be considered. The joint between semiconductor junction and case has a thermal resistance (T_{jc}). This data is available from the manufacturers but we can assume that for most devices it is approximately 2°C/W. Where the transistor is mounted to the heat sink, another thermal resistance exists (T_{mb}) and this can be assumed to be approximately 0.5°C/W.

The overall thermal resistance (R_{th}) required for the heat sink can be determined using:

$$R_{th} = \frac{T_j - T_a}{P} - (T_{jc} + T_{mb}) \text{ °C/W}$$

where

R_{th} = thermal resistance in °C/W
T_j = junction temperature (150°C)
T_a = ambient temperature
T_{jc} = thermal resistance between junction and case (2°C/W)
T_{mb} = thermal resistance between component and heat sink (0.5°C/W)

Let us use an *example* in order to use this equation.

A transistor is to be used as an amplifier and will generate 28 W. The surrounding air temperature will be 40°C. Calculate the thermal resistance and select an appropriate heat sink.

$$R_{th} = \frac{T_j - T_a}{P} - (T_{jc} + T_{mb}) \text{ °C/W}$$

$T_j = 150°C$
$T_a = 40°C$
$P = 28 \text{ W}$
$T_{jc} = 2°C/W$
$T_{mb} = 0.5°C/W$

$$R_{th} = \frac{150 - 40}{28} - (2 + 0.5) \text{ °C/W} = 1.43°C/W$$

A heat sink must now be chosen that has a *LOWER* thermal resistance than 1.43°C/W if the transistor is to be kept at a safe operating temperature, e.g. a 170-091 ($R_{th} = 1.1°C/W$) or 150-016 ($R_{th} = 1.0°C/W$) heat sink.

Power Transistor Selection

Power transistors are physically different from small signal devices because of the current demand placed upon them. Since most of the power is dissipated in the collector region this is where the heat is generated. It is usual for the collector region to be physically connected to the metal case or tag of the transistor package. This in turn will have holes drilled in it to allow the device to be bolted to a heat sink.

Note In some cases the transistor will have to be electrically isolated from the heat sink using an insulating washer (usually mica). This will have to be taken into account in the thermal resistance calculations.

Specifications

The maximum power that the device can handle is of prime importance and this in turn is a function of the collector current (I_C) and the collector-emitter voltage V_{CE} (as seen from the collector power curve of Fig. 3.5). A glance at the power transistor data sheets on page 146 will reveal the following specifications:

$I_C(AV)$ max — maximum average collector current (amperes)
P_{tot} — total permitted collector power (watts) usually quoted at a specific temperature.

P_C max.   — this means the same as P_{tot} and is an alternative specification preferred by some manufacturers.

V_{CEO} — maximum permissible collector-emitter voltage.

Power Transistor Selection Assignment

You have designed an amplifier to work at 38 V and draw a current of 2 A from its d.c. supply. A single n-p-n transistor is biased so that under quiescent conditions $I_C = 1.64$ A $V_{CE} = 25$ V. This delivers 35 W into its load when the output voltage is 20 V rms. Select a suitable transistor.

Method

1   Calculate the d.c. input power (P_{dc})
    $P_{dc} =$ d.c. power supply current $\times$ d.c. power supply voltage
    $= 2$ A $\times 38$ V $= 76$ W.
2   Calculate the a.c. output power (P_{ac})
    $P_{ac} = 35$ W.
3   Calculate the quiescent collector power (P_{CQ})
    $P_{CQ} = I_{CQ} \times V_{CEQ} = 1.64$ A $\times 25$ V $= 41$ W
    or since
    $P_{CQ} = P_{dc} - P_{ac} = 76$ W $- 35$ W $= 41$ W
    $P_{CQ} = 41$ W
4   Calculate the collector current I_C(rms)

$$I_C(\text{rms}) = \frac{P_{ac}}{V_{out}(\text{rms})} = \frac{35}{20} = 1.75 \text{ A}$$

$$I_C(\text{rms}) = 1.75 \text{ A}$$

The specification for the transistor can now be stated:

device type = n-p-n
$P_{tot} = 41$ W
$V_{CEQ} = 38$ V
$I_C(\text{rms}) = 1.75$ A

Any transistor with ratings higher than these requirements will be suitable; but remember the higher the ratings the more the device will cost so choose prudently, e.g. *TIP41*

$P_{tot} = 65$ W
$I_C(\text{AV}) = 6$ A
$V_{CEQ} = 40$ V.

The Class B Push-Pull Amplifier

For a number of reasons the single transistor power amplifier is very unsuitable for anything over about 4 watts. The main considerations being the power requirement of the device and the efficiency (or lack of it!) of the circuit. The input signal to an amplifier can be considered as swinging positive and negative about 0 V. We know that it is possible to bias a transistor at *cut-off* so that an input signal forces it into conduction. This is called Class B biasing.

It follows that an n-p-n device biased in Class B will be *turned on* by a positive going input signal while a p-n-p device in Class B will be *turned on* by a negative going input signal. It is sound reasoning then to use an n-p-n transistor to amplify the positive half cycles while a p-n-p device handles the negative half cycles as shown in Fig. 3.6.

Each transistor only handles 180° of input signal but by joining the two transistors together in the correct way a full cycle (360°) of output signal can be obtained. The circuit that does this is the Class B push-pull amplifier of Fig. 3.7.

When there is no input signal connected (under quiescent conditions) both T_1 and T_2 are off. So no current flows in the output — most efficient! As the input swings positive T_1 conducts and collector current (I_c) flows $\therefore I_e$ flows (since $I_e \simeq I_c$) through R_L developing a voltage (V_{out}) across it. When the input goes negative T_2 conducts I_c and I_e flows and current flows in R_L so V_{out} is developed across it. (Think of it as T_1 *pushing* the current through R_L while T_2 *pulls* it through R_L.)

The transistors complement each other in as much that when T_1 is on T_2 must be off because it is reverse biased, and vice versa. What a perfect solution to the power amplifier problems — or is it? Perform the associated practical investigation and see if you can discover the hiccup!

You will have found that the problem is due to the actual biasing. Both transistors are biased at *cut-off* which means the input signal must first overcome the base-emitter voltage (V_{BE}) to turn them on. This is about 0.6 V for a silicon device but the input characteristic is non-linear as shown in Fig. 3.8. Virtually no base current (I_B) will flow until V_{in} reaches about 0.6 V (Fig. 3.9) thus producing distortion in the output signal known as *crossover distortion*.

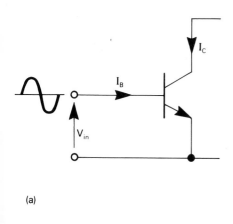

(a)

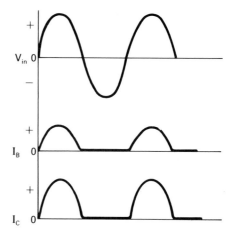

(b)

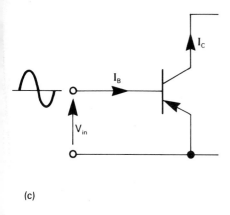

(c)

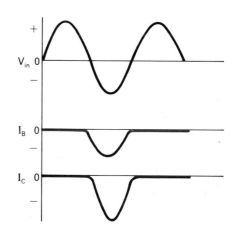

(d)

Fig. 3.6 (a) n-p-n transistor; (b) Signal waveforms; (c) p-n-p transistor; (d) Signal waveforms

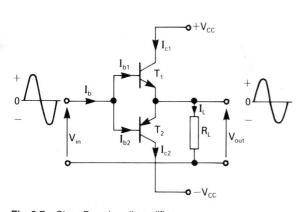

Fig. 3.7 Class B push-pull amplifier

Fig. 3.8 Input characteristic

PRACTICAL INVESTIGATION 3

The Class B Push-Pull Amplifier

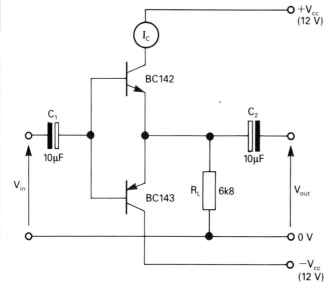

Equipment
CRO
Signal generator
Centre zero power supply (or two single power supplies)
Multimeter
BC142, BC143 transistors
$2 \times 10\ \mu F$ capacitors
6k8 resistor

Method
1 Build the circuit as shown ensuring that the power supplies are correctly connected.
2 Connect the signal generator and adjust to give a sine wave V_{in} of 3 V peak-to-peak at 1 kHz.
3 Monitor V_{in} and V_{out} with the CRO and sketch these on a common timescale.
4 Observe the output signal for values of V_{in} of 0.6 V, 1 V, 1.5 V, 2 V, 5 V and 8 V.
5 Draw a careful sketch of V_{in} and V_{out} when $V_{in} = 1.5$ V.
6 Disconnect the signal generator and measure the current flowing in the transistors (I_C)

Results
1 Explain why there is no output for low amplitude input signals.
2 When $V_{in} = 1.5$ V are both the positive and negative output half cycles the same amplitude?
— Justify your findings.
3 Try to suggest a way in which crossover distortion may be reduced.
4 What is the quiescent current flowing in the output transistors?

This non-linearity will be reflected in I_c and consequently V_{out}. The full effects are shown in Fig. 3.10.

This type of distortion occurs as one transistor *turns off* and the other *turns on*, in other words, as control *crosses over* from the n-p-n to the p-n-p device. Hence the term *crossover* distortion.

Reducing crossover distortion

Clearly, distortion of this type is unacceptable so how can it be reduced? Well it occurs because the transistors themselves are biased at cut-off and turning them on involves using the non-linear part of the input characteristic. The solution is to bias the transistors at the beginning of the linear

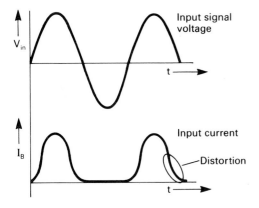

Fig. 3.9   Non-linear distortion

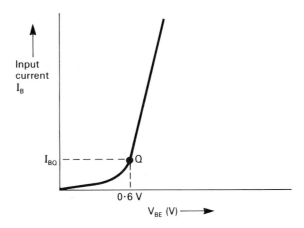

Fig. 3.11   Biasing beyond the non-linear region

part of the characteristic, i.e. apply about 0.6 V to the base of each device thus turning them on slightly.

THOUGHT

Surely if this is done the transistors are no longer in Class B bias. — True but nor are they in Class A bias, this is in fact Class AB biasing.

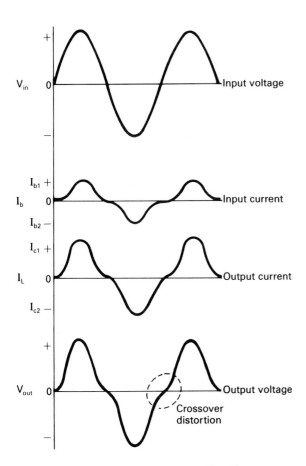

Fig. 3.10   Waveforms showing crossover distortion

Practical AB biasing

A potential divider arrangement could be used to bias both transistors into slight conduction. However a superior method employs two Si diodes as shown in Fig. 3.12.

The volt drop (0.6 V approx.) across D_1 forward biases T_1 slightly while the volt drop across D_2 forward biases T_2. In this way both transistors are turned on slightly so that the non-linear part of the input characteristic is overcome. The advantage offered by using diodes is that since they are manufactured from the same material as the transistor (silicon) they will have similar temperature characteristics. If the diodes are mounted on the same heat sink as the transistors they will undergo the same voltage change with temperature and the biasing will remain constant. Practical Investigation 4 gives you a chance to check the operation of this circuit for yourself.

An important point to note concerning any circuit that uses one device for each half cycle is the gain of each individual transistor. If there is the *slightest* difference in the h_{fe} of the two

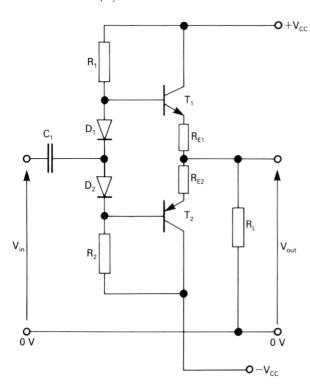

Fig. 3.12   Class AB biasing using Si diodes

transistors, one half cycle will be amplified more than the other. To prevent this, complementary pairs of power transistors (one p-n-p one n-p-n) are available as *matched pairs*, i.e. they have identical characteristics and so each produce exactly the same gain. When repairing such amplifiers, should one transistor be found to be faulty *both* must be replaced to ensure that correct operation is maintained.

From your deliberations you will have discovered that the power amplifier offers unity voltage gain, but can provide considerable current gain. It follows that the input signal to a power amp must itself be quite large and the output provided by transducers such as microphones, tape heads, pick up cartridges etc., is very small. Consequently such small inputs cannot successfully drive a power amplifier. Some form of small signal pre-amplifier will be required to provide a complete audio amplifier system. Investigation 5 extends the Class AB power amplifier so that it includes a small signal

driver and gives you the opportunity to perform a full test procedure and produce a detailed report.

Integrated Circuit Power Amplifiers

Today the integrated circuit (IC) or to use the media term *silicon chip* is very much an intrinsic part of electronics. In recent years miniaturization has moved forwards in leaps and bounds to provide us with smaller, neater and more attractive pieces of equipment. Integrated circuits seem capable of doing much more than their discrete circuit counterparts of only a few years ago, and now discrete component digital circuits are virtually unheard of for computers and logic control purposes. It perhaps is only natural that linear integrated circuits should, where possible, replace bulky discrete component circuitry in the world of analogue electronics. If you study any component catalogue you will see that it is possible to obtain complete circuits as an IC package, e.g. pre-amplifiers, r.f. amplifiers, voltage regulators, waveform generators and medium power amplifiers.

The convenience of using integrated circuits is self evident: quite complex circuits can be constructed by simply connecting a few additional components to a purpose-built integrated circuit. It is always wise, however, to consider the merits of both discrete and integrated circuits before accepting the IC as the universal solution to all electronic problems (Table 3.1).

Once this list has been analysed it can be seen that obviously the IC offers a number of advantages over the discrete circuit and that is why every week will see the introduction of new *varieties* of integrated circuits. Where there is a high current and power demand discrete circuits at this moment have the edge on the IC. It is probably this reason more than any other that restricts the availability and choice of IC amplifiers.

Common IC audio power amplifiers and their specification

LM380   — 14 pin package supplying 2 W into an 8 Ω load from a 20 V supply.

Class AB Push-Pull Amplifier

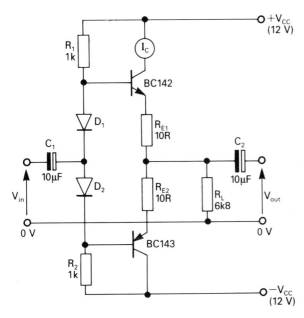

Equipment
CRO
Centre zero power supply
Signal generator
Multimeter
BC142, BC143 transistors
2 × silicon diodes
2 × 10R, 2 × 1k, 6k8 resistors
2 × 10 µF capacitors

Method
1   Build the circuit as shown.
2   Connect the signal generator and monitor V_{in} and V_{out} with the CRO.
3   Using a 1 kHz sine wave input signal, sketch the input and output waveforms when $V_{in} = 2$ V peak-to-peak.
4   Observing the output waveform carefully, carry out the following:
     (a)   short circuit D_1 (Bridge D_1 with a piece of wire)
     (b)   short circuit D_2
     (c)   short circuit D_1 and D_2
     (d)   restore diodes to their original state.
5   Disconnect V_{in} and measure the quiescent current (I_C) flowing in the transistors.

Results
1   What is the purpose of D_1 and D_2?
2   What is the quiescent current flowing in the circuit and how does this compare to the basic circuit of Investigation 3
3   What is the voltage gain of this amplifier?
4   Comment on the actual biasing of the transistors in this circuit.

PRACTICAL INVESTIGATION 5

Test and Report of Class AB Push-Pull Power Amplifier with Driver Stage

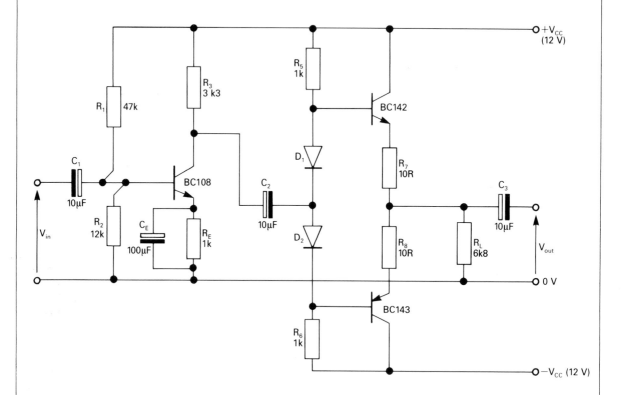

Equipment
Circuit and instruments from Investigation 4 plus: BC 108 transistor, 47 k, 12 k, 3k3, 1 k resistors, 10 µF, 100 µF capacitors.

Method
1 Connect the circuit as shown.
2 Determine the maximum and minimum input signal limits of the amplifier.
3 Produce a voltage gain/frequency response curve for the circuit.
4 Measure the quiescent current drawn from the power supply.

Results
These should be presented in the form of a concise report that includes the following:
1 Specification of the circuit (power-supply voltage, quiescent current, signal limits).
2 A fully labelled gain/frequency response curve.
3 Bandwidth information.
4 Suitable applications for the circuit.

Table 3.1   A comparison between the integrated and discrete circuit

ICs	Discrete
1 Miniaturization — large circuits in a small physical package.	Circuits bulky.
2 Reliability, due to consistency of manufacture.	Reliability determined by component type and circuit construction.
3 Common circuits cheap due to mass production techniques.	Individual components tend to cost more ∴ discrete circuits more expensive than equivalent IC.
4 Fault diagnosis relatively simple due to ease of access to pins of IC.	Fault diagnosis likely to require considerable skill for complex circuits.
5 Replacement usually easy due to plug-in ICs and sockets.	Dismantling involved — de-soldering etc.
6 A fault in any part of the IC renders it useless (this can prove to be expensive).	Fault finding possible down to a single component which can then be replaced.
7 Specialist circuits are very expensive due to limited production.	Any circuit can be tailor made.
8 Cooling the actual *silicon chip* inside the IC package is very difficult ∴ current and power limitations are placed on ICs.	Circuits of virtually any power requirements can be constructed and cooling provided.

TBA820m   — 8 pin package supplying 2 W into 8 Ω.

TDA2004   — dual Class B audio amp in an 11 lead tab. mounting package. Each amplifier is capable of providing 14 W into a 2 Ω load (9 W into 4 Ω).

TDA2030   — high power audio amplifier in a 5 lead package capable of supplying 21 W into 4 Ω.

Although it is fair to say that cost rises with power output, the price of such ICs is very reasonable when compared to an equivalent discrete circuit. This coupled with the very simple construction requirements makes them very suitable for a number of applications. We shall now look at the LM380 in detail.

The LM380 Audio Power Amplifier

General information

This is an audio amplifier with a fixed gain. Inputs to the circuit can be referenced to ground or a.c. coupled if required. The output signal is automatically centred at half the supply voltage. This amplifier has internal protection to limit the current and incorporates a thermal shutdown facility should overheating occur. It is supplied in a 14 pin package as shown in Fig. 3.13.

Note Pins 3, 4, 5 and 10, 11, 12 must be connected to ground and a heat sink approximately 6.4 cm^2 must be connected directly to the pins for cooling (it is usual to use a printed circuit board; in which case a 6.5 cm^2 area of copper can

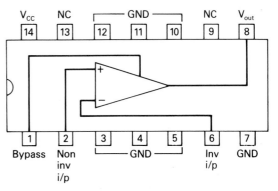

Fig. 3.13   LM380 integrated circuit power amplifier

PRACTICAL INVESTIGATION *6*

Test and Report on an IC Power Amplifier

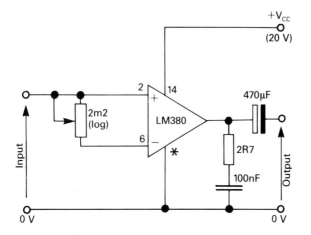

Equipment
CRO, signal generator
Power supply, multimeter
LM380
8 Ω loudspeaker
8 Ω 3 watt resistor
Crystal microphone
2M2 logarithmic potentiometer
2R7 resistor
470 μF, 100 nF capacitor

Method
1   Build the circuit as shown (observing the heat sink requirements).
2   Connect the power supply and adjust to give $V_{CC} = 20$ V.
3   Using the microphone test that the amplifier works and provides an audio output.
4   Turn off the power supply and replace the loudspeaker with an 8 Ω 3 watt resistor.
5   Connect the signal generator to provide the input signal, monitor V_{in} and V_{out} using the CRO.
6   Using an input signal frequency of 1 kHz *determine*:
     the voltage gain (A_v)
     the maximum output power
     (refer to the section on testing in Chapter 2 for help)
7   By taking such measurements as you require select a suitable frequency range and plot the gain/frequency response of the amplifier.
8   From your graph determine the bandwidth.

Results
Your results should be presented concisely in the form of a report that contains:
1   Details of the circuit.
2   A list of specifications.
3   The frequency response curve.
4   A comparison with the manufacturer's specification.
Any problems that were encountered and their solutions should also be fully documented.

be used for this purpose). Finally, integrated circuit amplifiers make fine pre-amplifiers that can be used to drive discrete circuit power amplifiers capable of providing very high audio outputs.

Typical Specifications

Fixed loop gain	= 50 (34 dB)
Supply voltage range	= 8 to 22 V max.
Quiescent current	= 7 mA
Input sensitivity	= 150 mV rms
Input resistance	= 150 kΩ
Load resistance	= 8 Ω
Bandwidth	= 100 kHz

Maximum Ratings

Supply voltage	= 22 V
Peak current	= 1.3 A
Input voltage	= ±0.5 V

This information shows that the amplifier would be very suitable for any low power audio application, but remember if it is to be supplied from a battery the current drain will exhaust it fairly quickly. Investigation 6 gives you a chance to build this amplifier and perform a test to compare the practical specifications with those quoted.

Power Amplifier Review

1 Power amplifiers are large signal amplifiers that use the greatest possible changes in voltage and current signals flowing in the amplifying device.

2 Single-ended amplifiers (one transistor) are biased in Class A.

3 Amplifier efficiency (η)

$$= \frac{\text{a.c. output power}}{\text{d.c. input power}} \times 100\%$$

4 The majority of losses in a transistor are collector power losses (P_C).

5 A power transistor generates heat that must be dissipated using a *heat sink*.

6 For improved efficiency two transistors biased in Class B push-pull mode can be used.

7 Class B produces crossover distortion that can be minimized by Class AB biasing.

8 Matched transistors are used for push-pull circuits in order that both positive and negative half cycles receive the same amplification.

9 Integrated circuit amplifiers are available for power applications up to about 20 W.

10 IC *pre-amplifiers* are often used to drive a discrete power output stage.

4

Feedback

In an amplifier or system, feedback is said to occur when a fraction of the output signal is taken and fed back into the input.

Feedback is something that occurs in everyday life and is in fact part of our natural control systems. When taking a shower, the output can be considered as the temperature of the water, our body senses this output and *feedback* occurs when we adjust the control that changes the temperature of the water. When driving a car there is constant feedback (via ourselves) between the car's direction (output) and the steering wheel position (input).

In electronics feedback can be deliberately applied in different amounts in order to make a system behave or perform in a desired fashion. To examine this concept more closely we will consider the amplifier since it is a fundamental building block of most systems. Study the familiar amplifier symbol shown in Fig. 4.1.

The voltage gain $A_v = \dfrac{V_{out}}{V_{in}}$

You can see from the diagram that there is no connection from the output back to the input. Feedback is not applied to this amplifier so it is said to be an *open loop* system. Now consider the following circuit.

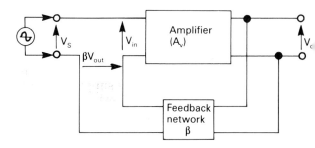

Fig. 4.2 Amplifier with feedback

The amplifier itself is the same but now a portion or fraction of the output signal is taken and fed back in series with the applied input signal. This will undoubtedly affect the performance of the amplifier and we need to predict how. A good starting point is to examine the terms involved and then derive the feedback equations.

A_v = Amplifier voltage gain without feedback.
V_{in} = Input voltage to amplifier.
V_{out} = Amplifier output voltage.
β = Feedback fraction.
V_s = Signal voltage applied to the whole circuit.

Feedback voltage

If β is the feedback fraction it follows that the feedback voltage itself will be $\beta \times V_{out} = \beta V_{out}$.

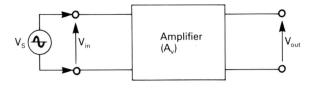

$$V_s = V_{in}$$

Fig. 4.1 Amplifier without feedback

Input signal V_{in}

The input signal to the system is V_s, but to this must be added the feedback voltage βV_{out}.

$$V_{in} = V_s + \beta V_{out}$$

This shows that the actual input signal (V_{in}) to a feedback amplifier is the sum of the input signal (V_s) and the feedback signal (βV_{out}). The feedback signal itself is derived from the output, clearly the affect that this will have on V_{in} will depend upon the phase of the feedback signal relative to the input signal (V_s). *Will V_{in} increase or decrease as the result of feedback?* If V_{in} increases due to βV_{out} then *positive feedback* is taking place. If V_{in} reduces courtesy of βV_{out} then negative feedback is occurring.

From this comes the realization that βV_{out} will in reality either be added *or* subtracted from V_s according to the nature of the feedback signal. If βV_{out} is in phase with V_s it will be added to V_s and the feedback is *positive*. If βV_{out} is in anti-phase with V_s it will be subtracted from V_s and the feedback will be *negative*.

Let us now continue with the derivation of the positive and negative feedback equations.

With reference to the block diagram of Fig. 4.2 we can show that:

$$V_{out} = A_v \times V_{in}$$

but V_{in} will be determined by the feedback voltage βV_{out}.

When βV_{out} is positive: $V_{in} = V_s + \beta V_{out}$
When βV_{out} is negative: $V_{in} = V_s - \beta V_{out}$

These expressions can now be substituted for V_{in} in the following equations as shown:

Positive feedback	*Negative feedback*
$V_{in} = V_s + \beta V_{out}$	$V_{in} = V_s - \beta V_{out}$
$V_{out} = A_v \times V_{in}$	$V_{out} = A_v \times V_{in}$
$\therefore \quad V_{out} = A_v(V_s + \beta V_{out})$	$\therefore \quad V_{out} = A_v(V_s - \beta V_{out})$

multiply out the brackets to give

$V_{out} = A_v V_s + A_v \beta V_{out}$	$V_{out} = A_v V_s - A_v \beta V_{out}$

Rearranging this gives

$V_{out} - A_v \beta V_{out} = A_v V_s$	$V_{out} + A_v \beta V_{out} = A_v V_s$

there are now *two* V_{out}s so factorizing gives

$V_{out}(1 - A_v \beta) = A_v V_s$	$V_{out}(1 + A_v \beta) = A_v V_s$

Rearrange to give V_{out}

$V_{out} = \dfrac{A_v V_s}{1 - A_v \beta}$	$V_{out} = \dfrac{A_v V_s}{1 + A_v \beta}$

Since V_s is the input signal to the whole circuit

$A_{pfb} = \dfrac{V_{out}}{V_s}$	$A_{nfb} = \dfrac{V_{out}}{V_s}$
$\dfrac{V_{out}}{V_s} = \dfrac{A_v}{1 - A_v \beta}$	$\dfrac{V_{out}}{V_s} = \dfrac{A_v}{1 + A_v \beta}$
$A_{pfb} = \dfrac{A_v}{1 - A_v \beta}$	$A_{nfb} = \dfrac{A_v}{1 + A_v \beta}$
Positive feedback equation	Negative feedback equation

Please note the following points carefully:

- Positive feedback results in an equation that has a *minus* sign in the denominator while the negative feedback equation has a *plus* sign in the denominator.
- There are two gain values:
  A_v which is the gain of the amplifier (or amplifying component) without feedback, often called the *open loop* gain (A_o).
  The overall gain of the complete amplifier circuit with feedback applied (A_{pfb}, A_{nfb}) is called the closed loop gain (A_c).

Effect of Positive Feedback (pfb)

If the feedback signal (βV_{out}) is in phase with the input signal (V_s) then positive feedback takes place, this will result in V_{out} increasing as shown in Fig. 4.3.

THOUGHT

If V_{out} increases won't βV_{out} increase? — Yes! And so V_{out} will increase more, resulting in a further increase in βV_{out} and so on. Instability may result from this condition.

Let us use a numerical example to help understand this.

Example 1
An amplifier has an open loop gain (A_v) of 25. Calculate the gain if 2% of positive feedback is applied.

$$A_v = 25 \qquad \beta = 2\% = 0.02$$

$$A_{pfb} = \frac{A_v}{1 - A_v\beta} = \frac{25}{1 - (25 \times 0.02)} = 50$$

So by introducing a mere 2% of positive feedback the gain of the system has doubled — this is wonderful! — *or is it?* Consider the same circuit with 4% positive feedback applied.

$$A_v = 25$$

$$\beta = 4\% = 0.04$$

$$A_{pfb} = \frac{25}{1 - (25 \times 0.04)} = \frac{25}{1 - 1} = \frac{25}{0} = \infty$$

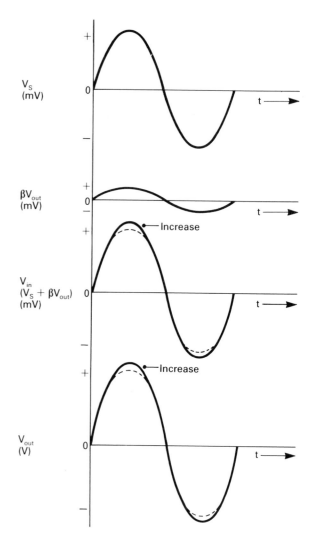

Fig. 4.3 Waveforms showing the effects of positive feedback

A_{pfb} is now infinity, i.e. very large!! Consequently the amplifier is now unstable. More about this later.

Negative Feedback (nfb)

If the feedback signal (βV_{out}) is in anti-phase (180° phase shift) to the input signal (V_s) then negative feedback takes place, this will result in a decrease in V_{out} as shown in Fig. 4.4.

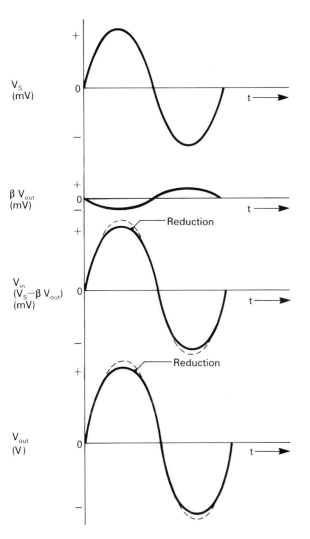

Figure labels (left axis):
V_s (mV)

βV_{out} (mV)

V_{in} $(V_s - \beta V_{out})$ (mV) — Reduction

V_{out} (V) — Reduction

Fig. 4.4 Waveforms showing the effects of negative feedback

THOUGHT

If V_{out} decreases βV_{out} will decrease and so eventually the output will fall to zero since it is the opposite to positive feedback — No! Think about this, βV_{out} causes V_{in} to decrease slightly so V_{out} decreases, this results in a decrease in βV_{out} which in turn allows V_{in} to rise slightly — so far from a cumulative reduction in V_{out}, negative feedback has a stabilizing effect on the gain, able to keep the gain reasonably constant despite variations in the components or external influences.

Let us use similar numerical values to show the result of negative feedback.

Example 2
An amplifier has an open loop gain $A_v = 25$. Calculate the gain if 1% of negative feedback is applied.

$$A_v = 25 \quad \beta = 1\% = 0.01$$

Negative feedback is applied, so the feedback signal (βV_{out}) will be subtracted from V_s hence reducing V_{in}. Let us use the negative feedback equation and see the change that this causes:

$$A_{nfb} = \frac{A_v}{1 + (A_v \beta)} = \frac{25}{1 + (25 \times 0.01)}$$

$$= \frac{25}{1 + 0.25} = 20$$

So the 1% nfb reduces the gain to 20.

Having considered these two equations it is quite acceptable to generalize and state that the general feedback equation is given by:

$$A_{fb} = \frac{A_v}{1 \pm (A_v \beta)}$$

By studying the foregoing examples it becomes plain that where amplifiers are concerned the thing to be avoided at all costs is positive feedback because this may make the amplifier unstable. If you are in any doubt that this is true consider the case of acoustic feedback that often occurs in public address systems. This is a perfect example of positive feedback in action.

If the gain of the amplifier in Fig. 4.5 is too high or the microphone is too close to the speaker the output from the speaker will be picked up by the microphone. The amplifier amplifies the *pick-up* which then emerges louder from the speaker to enter the microphone, etc. etc. The result is a dreadful howling that can usually be stopped only by turning the system off, and turning on again once the problem has been solved, i.e. different positioning of the microphone and/or reduced gain.

While this example considers airborne audio signals it can be seen that should any electrical signals find their way from output to input and cause positive feedback the result would be

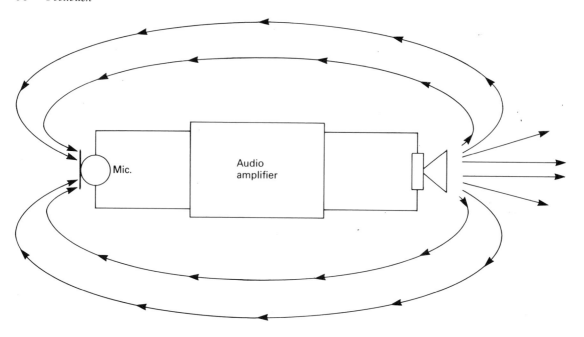

Fig. 4.5 Acoustic feedback

equally disastrous. This unwanted positive feedback may happen due to *invisible* paths from output to input via capacitive or electromagnetic coupling between stages, i.e. bad circuit layout or design. Negative feedback offers a number of advantages to a circuit that will now be considered, but please remember negative feedback must never be used to avoid instabilities due to bad circuit design.

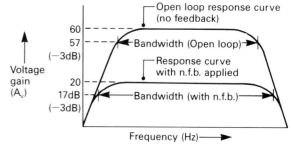

Fig. 4.6 Effects of negative feedback on gain and bandwidth

The Effects of Negative Feedback (nfb)

Bandwidth

Any change in the gain of an amplifier will alter its bandwidth as shown in Fig. 4.6.

You will remember the gain-bandwidth product from Chapter 2. From this any reduction in gain due to negative feedback, must result in a corresponding increase in the bandwidth since gain × bandwidth is a constant.

Gain stability

An amplifier built with brand new components may have a very high open loop gain (A_v) but this value could vary due to temperature changes, fluctuation in power supply voltage or the general effects of ageing. Any change in A_v will of course affect the value of V_{out}. This often becomes very obvious when a transistor fails and is replaced by one that has a higher or lower gain value. Ideally we would like A_v to remain constant or stable despite any variations in external conditions, but relying on an amplifier without negative feedback is likely to have disappointing results.

Example
An amplifier when first built has an open loop gain of 120, after a while due to ageing this falls to 90. Calculate the percentage reduction in gain. A_v (original) = 120, A_v (aged) = 90

$$\% \text{ reduction} = \frac{\text{original gain} - \text{new gain}}{\text{original gain}} \times 100\%$$

$$= \frac{120 - 90}{120} \times 100 = 25\%$$

So the gain has fallen by 25% — *not good!*
    If the same amplifier has 10% of negative feedback applied, calculate the gain with feedback before and after ageing. Determine the percentage reduction in gain and comment on the improvement resulting from negative feedback.

$$\beta = 10\% = 0.1$$

$$A_{nfb}(\text{original}) = \frac{120}{1 + (120 \times 0.1)} = \frac{120}{13} = 9.2$$

$$A_{nfb}(\text{aged}) = \frac{90}{1 + (90 \times 0.1)} = 9.0$$

$$\% \text{ reduction} = \left(\frac{9.2 - 9}{9.2}\right)100 = 2.17\%$$

So without negative feedback ageing caused a reduction in gain of 25% but with negative feedback applied the reduction in gain is only 2.17%.

THOUGHT _____

This is convincing but in the case of the open loop amplifier the gains are 120 and 90 but with nfb only 9.2 and 9.0, is this not rather a huge loss in gain? — Yes but it is all part of the trade off, gain is forsaken for gain stability. If a high gain is required this can be achieved by using more than one amplifying stage.

Very high gain amplifiers

There are many instances where the open loop gain of an amplifier is very high, i.e. 100 000 in a case like this:

$$A_{nfb} \simeq \frac{A_v}{\beta A_v} = \frac{1}{\beta} \qquad A_{nfb} \simeq \frac{1}{\beta}$$

This implies that the closed loop gain of a high gain amplifier with negative feedback is actually independent of the open loop gain (A_v) and is solely determined by the amount of feedback (β). *Provided of course that $\beta A \gg 1$.* Let us now use an example and both equations to prove this.

Example
An amplifier with an open loop gain (A_v) of 100 000 has 3% negative feedback applied. Calculate the closed loop gain.

$$A_v = 100\,000 \ (1 \times 10^5)$$

$$\beta = 3\% = 0.03$$

Using the full equation

$$A_{nfb} = \frac{A_v}{1 + (\beta A_v)} = \frac{1 \times 10^5}{1 + (0.03 \times 10^5)}$$

$$= \frac{1 \times 10^5}{3001} = 33.3$$

Using $A_{nfb} = \frac{1}{\beta} \simeq \frac{1}{0.03} = 33.3$

This is a very useful concept and one that you will meet again in the chapter on operational amplifiers.

Self Assessment 4

1 An inverting amplifier has an open loop gain of 140 and a bandwidth of 10 kHz. Calculate the gain and bandwidth resulting from the application of 4% negative feedback.
2 An amplifier has an open loop gain of 120. Calculate the amount of negative feedback required to reduce the gain to 50.

Distortion and noise

All amplifiers will introduce a certain amount of noise and distortion to any input signal. Negative feedback will reduce the gain of an amplifier so it must also reduce the amount of noise and distortion that the amplifier produces. If distortion (D) and noise (N) are taken to be voltages then it can be shown that:

$$D_{nfb} = \frac{D}{1 + (A_v\beta)} \qquad N_{nfb} = \frac{N}{1 + (A_v\beta)}$$

Example
An amplifier with an open loop gain of 120 has negative feedback applied. When providing an output of 10 V the amplifier introduces 20% of distortion and introduces 5 mV of noise. Calculate the improvement in distortion and noise when $\beta = 0.01$.

Open loop distortion = 20% of 10 V = 2 V

$$D_{nfb} = \frac{2}{1 + (120 \times 0.01)} = \frac{2}{2.2} = 0.91\ V$$

Noise

Open loop noise = 5 mV

$$N_{nfb} = \frac{5 \times 10^{-3}}{1 + (120 \times 0.01)} = 2.27\ mV$$

This shows that both noise and distortion are reduced by a factor of:

$$\frac{1}{1 + A_v\beta}$$

Remember that while this is true any subsequent stages will introduce additional noise so the overall noise level can be greater.

Input and output resistance (impedance)

The application of negative feedback involves a signal path from output to input and vice versa. This means that the input section and the output section of the amplifier are in fact connected together by the feedback. This will result in a change in both the input and output impedance of the amplifier. In order to examine the modifications that occur in impedances we need now to look at the various types of negative feedback and how they are applied.

Types of Negative Feedback

The signal that is *fed back* is obtained from the output signal, it can be derived from either the output voltage or the current flowing in the output stage. This feedback signal can then be applied either in series or in parallel (shunt) with the input signal. So we have a current or voltage derived feedback signal that can be series or shunt applied. There are consequently four types or methods of applying negative feedback.

Voltage-voltage feedback

This is sometimes called series voltage feedback, where the feedback is a *voltage* signal derived from the output *voltage* and applied to act in *series* opposition to the input signal as shown in Fig. 4.7.

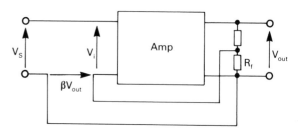

Fig. 4.7   Voltage-voltage (series voltage) feedback

The feedback voltage (βV_{out}) is derived from the output voltage (V_{out}) and is fed back to act in series with the input voltage signal V_s. A simple way of obtaining V_{out} is to use a potential divider. A practical example of this is the emitter or source follower circuits.

Note   In this circuit the output is taken across the emitter and source resistors.

Here the feedback voltage $\beta V_{out} = V_{out}$; this means that $\beta = 100\%$ thus the emitter and

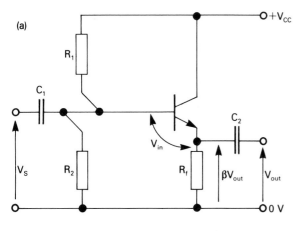

(a)

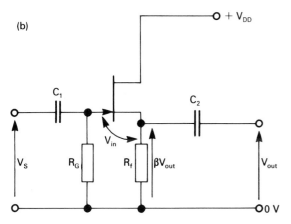

(b)

Fig. 4.8 (a) Emitter follower circuit; (b) Source follower circuit

source follower will have unity voltage gain, i.e. if $A_v = 150$

then $A_{nfb} = \dfrac{A_v}{1 + (A_v\beta)} = \dfrac{150}{1 + (150 \times 1)}$

$= \dfrac{150}{151} = 0.99$

$\therefore A_{nfb} \approx 1$.

The input impedance will be increased by series voltage feedback while the output impedance will decrease.

Voltage-current feedback

Also known as series current feedback. The feedback signal is a *voltage* developed from the

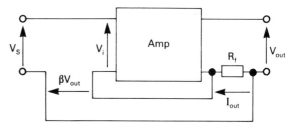

Fig. 4.9 Voltage-current (series current) feedback

output *current*, i.e. current derived and applied to act in *series opposition* to the input signal as shown in Fig. 4.9.

V_{out} is derived from I_{out} and is fed back to act in series with the input voltage signal V_s. The common emitter and common source amplifiers Figs 4.10 and 4.11 are practical examples of series current feedback, where the emitter and source resistors are undecoupled.

Series current feedback increases the input and output impedance. It is worth noting that the two types of voltage feedback will affect the voltage gain (A_v) of the amplifier while leaving the current gain (A_i) unchanged.

Before moving on to current feedback it is worth summarizing the effects of series voltage and series current feedback which are of course *both* examples of *voltage* feedback (Table 4.1).

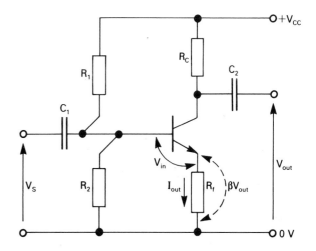

Fig. 4.10 Common emitter circuit

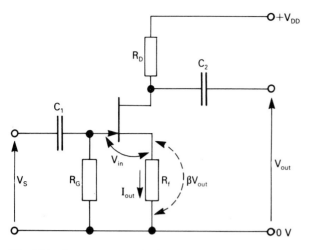

Fig. 4.11   Common source circuit

Here the feedback signal is the *current* βI_{out} which is derived from V_{out} and fed back to act in parallel to the input signal V_s. This type of feedback can be used as shown by the circuit in Fig. 4.13.

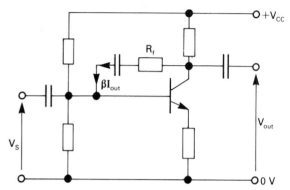

Fig. 4.13   Current-voltage feedback

Now perform the two Practical Investigations 7 and 8 which help to highlight the results of negative feedback.

Current-voltage feedback

Also called shunt voltage feedback. The feedback signal is a *current* signal developed from the output *voltage* and applied to act in parallel (shunt) opposition to the input signal (Fig. 4.12).

Shunt voltage feedback reduces both the input and output impedance. This technique is often used in single-stage audio amplifiers to provide a selective gain or as a tone control circuit.

Current-current feedback

Alternatively known as shunt current feedback. The feedback signal is a *current* developed from the output *current* and applied to act in parallel opposition to the input signal (Fig. 4.14).

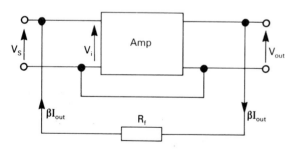

Fig. 4.12   Current-voltage (shunt voltage) feedback

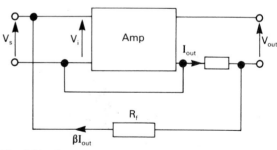

Fig. 4.14   Current-current (shunt current) feedback

Table 4.1   Voltage feedback

Type	Voltage Gain (A_v)	Input Impedance (Z_{in})	Output Impedance (Z_{out})
Series voltage	Reduced	Increased	Decreased
Series current	Reduced	Increased	Increased

PRACTICAL INVESTIGATION **7**

Negative Feedback (Series Voltage)
The Emitter Follower

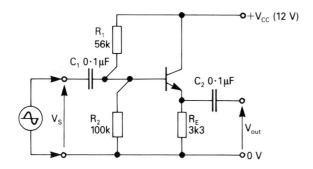

Equipment
Power supply
Signal generator
Oscilloscope
n-p-n general-purpose transistor
56 k, 100 k, 3k3 resistors
2×0.1 μF capacitors
Decade resistance box
6 cycle log/lin graph paper

Method
1   Build the circuit as shown.
2   Monitor V_s and V_{out} and adjust the signal generator to give a 1 kHz undistorted sine wave at the output using the procedures outlined in Chapter 2.

3   Calculate the voltage gain A_v using $A_v = \dfrac{V_{out}}{V_s}$.

4   Comment on the phase relationship between input and output signals.
5   Measure the input and output impedance of the amplifier.
6   Produce a frequency response curve for the amplifier showing how the voltage gain (A_v) in dB varies with frequency.
7   From your graph estimate the bandwidth of the amplifier.

The feedback current βI_{out} is derived from the output current I_{out} and acts in parallel to the input signal V_s. A practical example of this type of feedback is shown in Fig. 4.15.

Shunt current feedback reduces the input impedance while increasing the output impe-dance. Table 4.2 indicates the effects of shunt voltage and shunt current feedback which are both examples of *current* feedback. This type of feedback is often used to control the gain over several stages in an amplifier system.

Table 4.2 Current feedback

Type	Current Gain (A_i)	Input Impedance (Z_{in})	Output Impedance (Z_{out})
Shunt voltage	Decreased	Decreased	Decreased
Shunt current	Decreased	Decreased	Increased

PRACTICAL INVESTIGATION *8*

Negative Feedback (Series Current)
The Common Emitter Amplifier

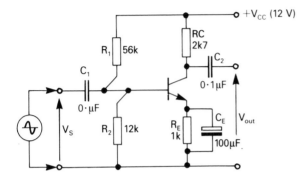

Equipment
n-p-n general-purpose transistor
56 k, 12 k, 2k7, 1 k resistors
Two 0.1 μF capacitors
100 μF capacitor
Dual beam oscilloscope (CRO)
Power supply
Signal generator
Decade resistance box
6 cycle log/lin graph paper

Method
1   Build the circuit shown above.
2   Connect the power supply and signal generator. Monitor V_s and V_{out} with the CRO and check that the circuit amplifies.
3   Set the signal generator to give a 1 kHz sine wave V_{out} that is undistorted.
4   Using the procedures outlined in Chapter 2 carry out the following tests and record the results.
   (a)   Voltage gain (A_v)
   (b)   Frequency response (plotted as A_v in dB against frequency)
   (c)   Input impedance
   (d)   Output impedance (*this must be done very carefully!*)
   From your frequency response plot, determine the bandwidth (B).
5   Disconnect C_E so that a.c. negative feedback occurs and repeat the tests plotting the frequency response curve on the same axis.

Results
Compare the results from both tests and verify that the statements previously made about the effects of negative feedback are true in practice.

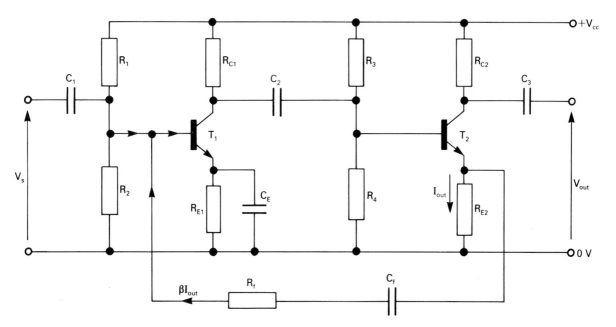

Fig. 4.15 Shunt current feedback in a two stage amp

A.C. and D.C. Feedback

It is possible to employ a.c. or d.c. negative feedback or both as the occasion demands. This is usually achieved by the prudent use of capacitors. You will remember from your studies of the common emitter and common source amplifiers that d.c. negative feedback is required for thermal stability while a.c. negative feedback is prevented either fully or partly using a bypass capacitor as shown in Figs 4.16 and 4.17.

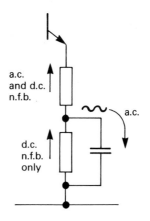

Fig. 4.17 Partial a.c. negative feedback

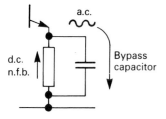

Fig. 4.16 d.c. negative feedback only

Likewise if only a.c. negative feedback is required the inclusion of a series blocking capacitor prevents d.c. negative feedback Fig. 4.18.

It is worth remembering that an amplifier or multi-stage circuit may employ more than one

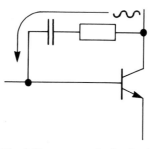

Fig. 4.18 a.c. negative feedback only

type of negative feedback and that during fault finding the existence of any feedback path will affect the stages that are linked together by it.

Feedback Review

1   Feedback occurs when a portion or fraction of a system's output is fed back to its input.
2   There are two main forms of feedback — positive and negative feedback.
3   Positive feedback occurs when the feedback signal is in phase with the input signal.
4   Negative feedback occurs when the feedback signal is in anti-phase to the input signal.
5   The general feedback equation is:

$$A_{fb} = \frac{A_v}{1 \pm (A_v \beta)}.$$

6   The positive feedback equation is:

$$A_{pfb} = \frac{A_v}{1 - (A_v \beta)}.$$

7   The negative feedback equation is:

$$A_{nfb} = \frac{A_v}{1 + (A_v \beta)}.$$

8   Positive feedback may result in instability and is therefore highly undesirable in amplifiers.
9   Negative feedback results in improved gain stability and consequently is usually employed in amplifiers.
10   Negative feedback has the following effects on an amplifier:
   (a)   Reduces the gain.
   (b)   Increases the bandwidth.
   (c)   Improves the gain-stability.
   (d)   Reduces noise and distortion.
   (e)   Modifies the input and output impedance.
11   If the gain of an amplifier is very high the closed loop gain can be approximated using:

$$A_{nfb} \simeq \frac{1}{\beta}$$

12   There are four types of negative feedback.
   ● Series voltage feedback (voltage-voltage feedback). The feedback is a voltage signal derived from the output voltage to act in series opposition to the input signal.
   ● Series current feedback (voltage-current feedback). The feedback is a voltage signal derived from the output current to act in series opposition to the input signal.
   ● Shunt voltage feedback (current-voltage feedback). The feedback is a current signal derived from the output voltage to act in parallel opposition to the input signal.
   ● Shunt current feedback (current-current feedback). The feedback is a current signal derived from the output current to act in parallel opposition to the input signal.
13   Feedback may be a.c., d.c., or both.

Self Assessment Answers

Self Assessment 4

1   $A_{nfb} = \dfrac{A_v}{1 + (A_v)} = \dfrac{140}{1 + (140 \times 0.04)} = 21.2$

Bandwidth

$B = 10\,\text{kHz} \quad A_v = 140$

Gain bandwidth product $= GB = 140 \times 10 \times 10^3 = 1.6 \times 10^6\,\text{Hz}$

$\therefore$ if $A_{nfb} = 21.2$, $B = \dfrac{GB}{A_{nfb}}$

$= \dfrac{1.6 \times 10^6}{21.2} = 66\,\text{kHz}$

2   $A_{nfb} = \dfrac{A_v}{1 + (A_v \beta)}$

$A_v = 120 \quad A_{nfb} = 50 \quad \beta = ?$

The formula must be transposed to find β.

$50 = \dfrac{120}{1 + (A_v \beta)} \quad \therefore \quad \dfrac{120}{50} = (1 + A_v \beta)$

$(1 + A_v \beta) = 2.4$

$\therefore A_v \beta = 2.4 - 1 \quad A_v \beta = 1.4$

$A_v = 120 \therefore \beta = \dfrac{1.4}{120} = 0.012 = 1.2\%$

5

The Operational Amplifier

The operational amplifier is a differential amplifier with an extremely high voltage gain. Known as the *Op-Amp*, this is a very interesting integrated circuit amplifier that was originally developed for use in analogue computers. Here they were connected to make circuits that would perform mathematical operations including summing, multiplication and integration. The original op-amps were quite bulky since they were constructed using discrete components that once built were *potted* making a black box with terminals or pins for connections.

Once miniaturization became commonplace these amplifiers were produced in integrated circuit form. Although originally very expensive, mass production methods have now made the common types very cheap indeed, and today they form a very useful building block for a vast range of electronic circuits.

What exactly is the 'op-amp'?

There are many types available but perhaps the most common one is the 741, this is available as an 8 pin dual-in-line (dil) package or as a fourteen pin package containing two devices (code number 747). Since all op-amps behave in a similar fashion, we shall concentrate on the 741 type for simplicity.

Construction

The 741 contains 20 transistors, 11 resistors and a capacitor, all fabricated on a single chip of silicon (Fig. 5.1).

This impressive list of internal components making up the op-amp can now be completely forgotten. It is not necessary to consider the actual circuit, only its performance as a building block is important. The device is represented by the circuit symbol shown in Fig. 5.2.

You can see from the pin connection diagram (Fig. 5.3) how the circuit symbol relates to the practical 8 pin integrated circuit. Note the notch (or in some cases a circle) at one end of the IC. With this pointing away from you numbering starts with the top left-hand pin (marked sometimes with a spot) and proceeds anticlockwise as shown.

You will notice from the connection diagram that there are pins that obviously do something but do not appear to be shown connected to the circuit symbol. This is a convention that has been adopted to keep diagrams simple; you will often find that power supply connections on diagrams are not shown, they are just assumed to be there!

Operation

In order to keep life as simple as possible we shall first consider the basic operation of the op-amp before attempting a detailed examination of its various specifications and parameters. A glance at the circuit symbol for the device shows that it is an amplifier that has a single output but two inputs. These inputs are labelled inverting ($-$) and non-inverting ($+$). From amplifier knowledge already gained you should be aware that a signal applied to the inverting input will produce an anti-phase output, while a signal applied to the non-inverting input will produce an output with zero phase shift.

It was stated in the opening passage that the op-amp is a differential amplifier. As it has two inputs it will produce an output that is an amplified version of the difference between the

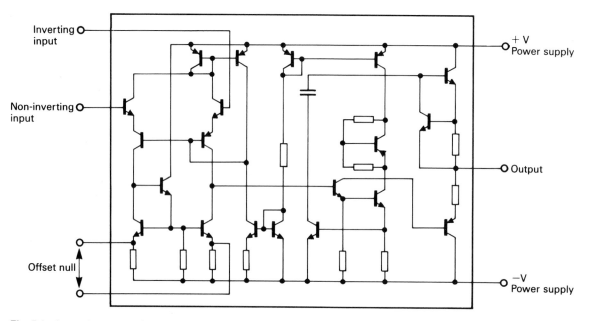

Fig. 5.1   Internal structure of the 741

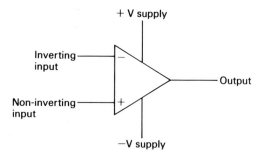

Fig. 5.2   Operational amplifier circuit symbol

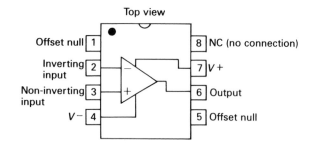

two input signals, i.e. if the signal applied to the (−) input is 25 mV and the signal applied to the (+) input is 18 mV the output signal will be

$$V_{out} = 25\,mV - 18\,mV \times A_v = -7\,mV \times A_v$$

(Minus because the input to the inverting terminal is greater than the input to the non-inverting terminal.)

THOUGHT

So if both input signals had the same amplitude the output would be zero? — Precisely! If the input signals are exactly the same they cancel ∴ the output will be an amplified version of zero.

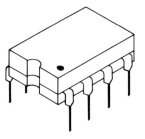

Fig. 5.3   Packaging and pin connections

If one of the inputs is connected to the signal earth or ground line then the device becomes a single input amplifier that can be either inverting or non-inverting according to how it is connected (Figs 5.4 and 5.5).

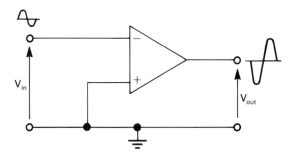

Fig. 5.4 Inverting amplifier

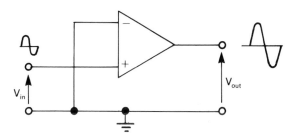

Fig. 5.5 Non-inverting amplifier

An important point to be aware of is that the op-amp is direct-coupled throughout, this means that inside the integrated circuit no capacitors are included in the signal path, therefore the device will amplify d.c. signals extremely well.

The story so far is that the op-amp can be connected to act as an inverting, non-inverting and differential amplifier that will amplify d.c. and a.c. signals.

Operational Amplifier Characteristics

The characteristics of a device indicate its practical performance. As with all things, e.g. a car, there will be the *ideal* characteristics which reflect the performance we desire in a perfect world and the practical or actual characteristics that exist in reality.

Example
The ideal characteristics for a car might be:

0–60 mph in zero seconds, maximum speed = infinite, fuel economy (miles per gallon) = infinite.

The actual characteristics are likely to be more modest:

0–60 mph in 8 seconds, maximum speed = 120 mph, fuel economy = 20 mpg at 56 mph.

A similar situation exists for the operational amplifier with an *ideal* device having the following characteristics:

Voltage gain (A_v)	– infinite
Input resistance	– infinite
Output resistance	– zero
Bandwidth	– infinite

In practice the actual characteristics for a 741 are:

Voltage gain (A_v)	– 106 dB (numerical gain = 200×10^3)
Input resistance	– 1 MΩ
Output resistance	– 75 Ω
Bandwidth	– up to 1 MHz

These specifications do not help us to see how the device can be used in a circuit. To do this a close look must be taken at the two really crucial parameters: the voltage gain (A_v) and the bandwidth (B) since they are interrelated.

The voltage gain (A_v)

This is extremely high, with 106 dB (200×10^3) quoted as the maximum gain – but this gain is not the same at all frequencies! In fact the low frequency gain is very high and falls rapidly as the frequency increases (as shown in Fig. 5.6).

This fall off in gain occurs at a constant rate such that at a frequency of about 1 MHz the gain for a 741 is 0 dB (unity).

Even at low frequencies the enormous gain is not really much help since the smallest input signal will result in the output immediately saturating at the level of the power supply voltage. This produces a switching rather than an amplifying action (which may be useful in

---

PRACTICAL INVESTIGATION 9

The Operational Amplifier Open-Loop Gain

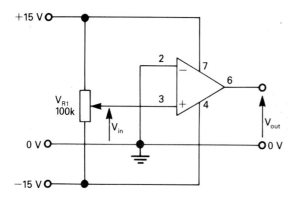

Equipment
Dual power supply
(or two single supplies)
741 op-amp
Dual beam oscilloscope (CRO)
100 kΩ potentiometer

Method
1   Connect the circuit as shown making sure that the centre or earth terminal of the supply forms the 0 V rail of the circuit.
2   Monitor V_{in} and V_{out} with the CRO switched to d.c., i.e. the a.c./d.c. switch on both channels in the d.c. position.
3   Set V_{R1} to give $V_{in} = 0$ V (this will be approximately mid-range).
4   Observe V_{out} as V_{R1} is varied about the centre position to give a positive and negative V_{in}.

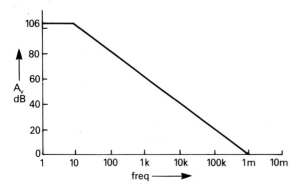
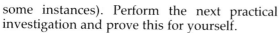

Fig. 5.6   741 gain/frequency characteristic

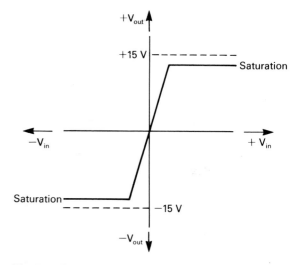

Fig. 5.7   Open loop response

some instances). Perform the next practical investigation and prove this for yourself.

If you were to draw a graph showing the response of the open loop op-amp it would probably look like Fig. 5.7.

This shows clearly that the tiniest input signal results in the op-amp being driven straight into saturation due to its very high gain. To operate the device effectively as an amplifier it is necessary to use the principles of negative feedback to reduce the gain to the required level. Your knowledge of feedback will tell you that once negative feedback is applied to an amplifier the following things occur:

1   The voltage gain is reduced.
2   The bandwidth widens.
3   The input and output resistances are modified.
4   The amplifier gain stability is improved.

Bandwidth (*B*)

We know that voltage gain and bandwidth are linked together, this is very clearly indicated by the frequency response curve of Fig. 5.6. Alter the voltage gain (A_v) using negative feedback and the bandwidth will change.

This is another example of the gain-bandwidth product (*GB*) in action.

Remember Gain-bandwidth product (*GB*) = numerical value of voltage gain (A_v) × bandwidth (*B*) in Hz.

Consider once more the 741 frequency response curve shown in Fig. 5.8.

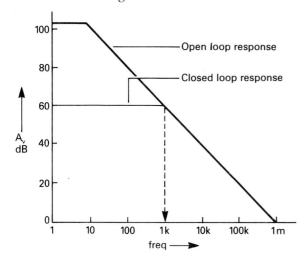

Fig. 5.8   Op-amp frequency response

The gain of the op-amp at 1 MHz is unity (0 dB).

$G = 1$ $B = 1$ MHz

Gain-bandwidth product $GB = 1 \times 1 \times 10^6 = 1$ MHz

$GB = 1$ MHz *this is a constant!*

Given the gain-bandwidth product for an amplifier the voltage gain or the bandwidth can be estimated.

Example
Supposing the gain (A_v) of the amplifier is reduced by the application of negative feedback to 60 dB — determine the bandwidth.

$A_v = 60$ dB ∴ numerical value

$$= A.\log\left(\frac{60}{20}\right) = 1000$$

So $G = 1000$ $GB = 1$ MHz since $GB = G \times B$

$$B = \frac{GB}{G} = \frac{1 \times 10^6}{1000} = 1000$$

So when $A_v = 60$ dB $B = 1$ kHz

Likewise if a bandwidth of 10 kHz is desired the gain required to produce this can be found in a similar way:

$GB = 1$ MHz $B = 10$ kHz

$$G = \frac{GB}{B} = \frac{1 \times 10^6}{10 \times 10^3} = 100$$

So to provide a 10 kHz bandwidth, negative feedback must be applied so that A_v is reduced to 100.

in dB $A_v = 20$ Log $100 = 40$ dB

By using an op-amp life is simpler. Once the gain-bandwidth product has been found the voltage gain and hence the bandwidth can be determined by applying negative feedback.

THOUGHT

This is fine if you have the frequency response curve of an op-amp but what if this is not available? Data sheets and catalogues contain detailed specifications relating to the various types of operational amplifiers, see pages 154 and 155. One of the details quoted is the transition frequency (f_T) or the unity gain bandwidth.

This is the frequency at which the gain has fallen to unity (0 dB) and is in fact the upper frequency limit of the amplifier. Since the lower frequency limit is 0 Hz it also represents the bandwidth (*B*). The maximum or typical voltage gain is also given and so the gain-bandwidth product can be determined.

Practical Operational Amplifier Circuits

The operational amplifier is used as a feedback amplifier. The design of circuits can be made simple by treating the device as *ideal* for the purposes of analysing circuit performance. This is summed up by utilizing the following rules:

1 The input current ($I_{i/p}$) drawn by the inverting (+) and non-inverting (−) terminals is zero!
2 The voltage dropped (developed) across the input terminals (V_d) is zero!
3 The open loop gain of the device is infinite!

The inverting amplifier

A typical circuit is shown in Fig. 5.9.

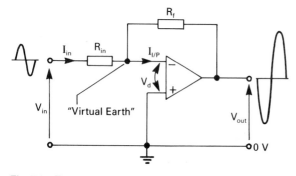

Fig. 5.9   The inverting amplifier

Since the amplifier is inverting, the output will be in anti-phase to the input. To gain an understanding of this mode of operation we must first consider the *virtual earth* concept.

The Virtual Earth

The input current to the device ($I_{i/p}$) is considered to be zero (*Rule (1)*).

The voltage dropped between the inverting and non-inverting terminals (V_d) is also zero (*Rule (2)*). The non-inverting terminal (+) is connected to the earth or 0 V line and if these two rules are observed then the voltage dropped between the two input terminals (V_d) is zero! (As it must be if zero current is passing.) Since the (+) input is earthed the (−) input must also be at earth potential as well! But due to the fact that the op-amp is not *ideal* there will be a tiny input current ($I_{i/p}$) and a very small volt drop (V_d), so the non-inverting terminal will *virtually be at earth potential*.

Well this is fine but it does not actually explain the operation of the circuit yet. The circuit consists of a very high gain amplifier that has an input resistor R_{in} and a feedback resistor R_f. R_f is connected between the inverting input terminal and the output terminal. The amplifier is inverting so the resistor R_f is applying negative feedback and will consequently determine the gain.

Now the output signal ($-V_{out}$) is fed back to the inverting terminal which we know to be *virtually at earth*. Consequently the full input voltage (V_{in}) will be dropped across R_{in}

$$\therefore I_{in} = \frac{V_{in}}{R_{in}}$$

This input current I_{in} will not flow into the op-amp since we are assuming it draws zero current! ($I_{i/p} = 0$.) *Where will it flow?* Through R_f!

$$\therefore I_{in} = I_f$$

So $\dfrac{-V_{out}}{R_f} = \dfrac{V_{in}}{R_{in}}$ (minus sign indicates phase inversion)

$$A_v = \frac{-V_{out}}{V_{in}} = \frac{-R_f}{R_{in}}$$ Where A_v is the closed loop gain of the circuit.

THOUGHT

What about the gain of the op-amp itself? This does not appear in the equations! — No it does not. This is because the op-amp's gain is assumed to be infinite. It is thus the ratio of the two resistors that determines the closed loop gain of the circuit.

Design of an inverting amplifier now becomes simplicity itself.

Example
A 741 is to be used as an inverting amplifier. Determine the resistor values required to provide a voltage gain of −10 and calculate the resulting bandwidth.

1 $-A_v = \dfrac{R_f}{R_{in}}$ $-10 = \dfrac{R_f}{R_{in}}$

2 *Decide on a value for R_{in}* Theoretically this could be any value but remember if R_{in} is chosen to be low this will draw too much current from whatever is supplying the input signal into the circuit since R_{in} = input resistance of the circuit. Therefore make R_{in} a reasonably high value, e.g. 10 k.
So $R_{in} = 10\,k$

3 *Determine the value of R_f*

$-A_v = \dfrac{R_f}{R_{in}}$ $\therefore R_f = -A_v \times R_{in}$

$= -10 \times 10 \times 10^3 = 100\,k$

$R_f = 100\,k$

4 *Calculate the bandwidth*
The frequency at which A_v = unity is 1 MHz for the 741.
Gain-bandwidth product = GB
$= 1 \times 1 \times 10^6 = 1\,MHz$
the Gain of the inverting amplifier is now 10.

$B = \dfrac{GB}{G} = \dfrac{1 \times 10^6}{10} = 100\,kHz$

So the bandwidth is 100 kHz. This will be the upper frequency limit since the lower frequency limit is d.c. (0 Hz). The inverting amplifier investigation helps to illustrate these points.

If you have performed the investigation successfully I hope you have made the following discoveries:

1 The amplifier's gain and frequency response is similar to the theoretical prediction Fig. 5.8. (Page 71.)
2 The gain-bandwidth product is proved to be a constant (making allowances for experimental errors!)
3 The gain falls at the rate of approximately 20

decibels per decade (6 dB/octave) once the −3 dB frequency has been reached, i.e. A_v at 10 kHz = 40 dB, A_v at   100 kHz = 20 dB (from 10 kHz to 100 kHz is one decade).

If the −3 dB frequency occurs at 10 kHz giving A_v = 40 dB, then the gain at 20 kHz will be 34 dB. One octave is a doubling of frequency. The gain, therefore, falls at 6 dB per octave.

The non-inverting amplifier (see Fig. 5.10)

As expected the input signal is applied to the non-inverting terminal but note that the input resistor and feedback resistor are connected to the inverting input. This may seem strange but it has to be like this because the feedback is always negative and is used to determine the gain. The inverting input is still the *virtual earth* point due to the high gain of the op-amp and the golden rules (1) and (2) relating to zero input voltage (V_d) and current ($I_{i/p}$) signals.

The circuit operates in the following way. The input signal V_{in} is applied to the non-inverting terminal and because there is zero volt drop between the input terminals V_{in} is applied across R_{in} so:

$I_{in} = \dfrac{V_{in}}{R_{in}}$

R_f and R_{in} make up a potential divider with V_{in} applied across R_{in}. V_{out} is applied across $R_f + R_{in}$

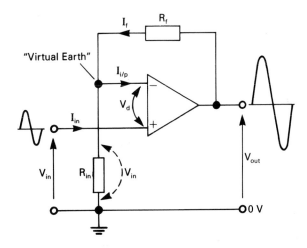

Fig. 5.10   Non-inverting amplifier

PRACTICAL INVESTIGATION *10*

The Op-Amp as an Inverting Amplifier

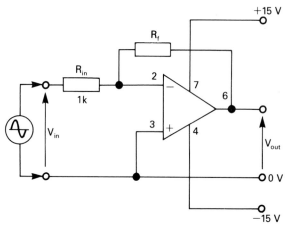

Equipment
Dual power supply (or two single supplies)
Oscilloscope (CRO)
741 Op-amp
Breadboard
Signal generator
6 cycle log/line graph paper
2 × 1 k resistors 10 k, 100 k

Method
1   Build the circuit as shown above to give $A_v = -100$ $(R_f = 100 \text{ k})$.
2   Monitor V_{in} and V_{out} with the CRO and check that the gain is as designed using a 1 kHz sine wave input signal of 100 mV peak-to-peak.
3   Set the signal generator to a low frequency (10 Hz) and monitor V_{out}, increase the frequency slowly and record the -3 dB frequency, i.e. frequency at which

$$A_v = 0.707 \times \frac{V_{out}}{V_{in}}.$$

4   Repeat the above steps with R_f changed to give gains of -10 (20 dB) and unity (0 dB) respectively.

Results
1   Plot, on the same axes the frequency response curve (A_v in dB against frequency) of each amplifier.
2   For each amplifier calculate the gain-bandwidth product.
3   Using your graph determine the rate at which the gain falls as the frequency increases.

and so using normal potential divider circuit theory:

$$V_{in} = \frac{R_{in}}{R_f + R_{in}} \times V_{out} \qquad A_v = \frac{V_{out}}{V_{in}}$$

so $\dfrac{V_{out}}{V_{in}} = \dfrac{R_{in} + R_f}{R_{in}} = \dfrac{R_f}{R_{in}} + \dfrac{R_{in}}{R_{in}} = \dfrac{R_f}{R_{in}} + 1$

$$A_v = \left(1 + \frac{R_f}{R_{in}}\right)$$

for the non-inverting amplifier.

This circuit has the added advantage of having a very high input resistance.

The unity gain buffer

You will recall from the section on amplifiers that there is often a need in electronics for matching a high impedance to a low impedance without introducing any gain or phase shift. The op-amp is extremely useful in providing this facility.

PRACTICAL INVESTIGATION **11**

The Op-Amp as a Non-inverting Amplifier

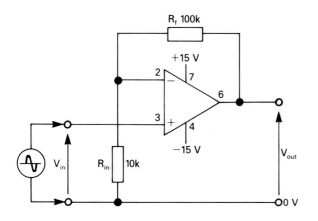

Equipment
Dual power supply
Oscilloscope (CRO)
741 op-amp
Breadboard
Signal generator
10 k, 100 k resistor

Method
1. Build the circuit as shown above.
2. Connect the signal generator and monitor V_{in} and V_{out} using the CRO.
3. Using a 1 kHz, 100 mV peak-to-peak sine wave input signal verify the formula.

$$A_v = \left(1 + \frac{R_f}{R_{in}}\right).$$

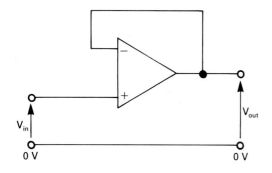

Fig. 5.11 The op-amp as a unity gain buffer

From Fig. 5.11 it can be seen that the negative feedback is 100% since all of V_{out} appears at the inverting input.

$$\therefore \quad V_{out} = V_{in}$$

$$A_v = \frac{V_{out}}{V_{in}}$$

$$\therefore \quad A_v = 1 \text{ (unity)}$$

This simple circuit is an excellent buffer that uses no external components at all!

The summing amplifier

This is a special version of the inverting amplifier.

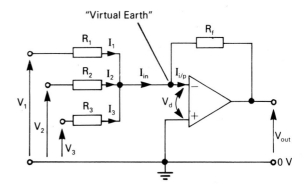

Fig. 5.12 The op-amp as a summing amplifier

Study Fig. 5.12 noting that again the *virtual earth* concept applies, and that $I_{i/p} = 0$ and $V_d = 0$. With the summing amplifier the output (V_{out}) is the sum of the applied input voltages multiplied by the gain of the amplifier.

The input current $I_{in} = I_1 + I_2 + I_3$

$$I_1 = \frac{V_1}{R_1}, \quad I_2 = \frac{V_2}{R_2}, \quad I_3 = \frac{V_3}{R_3}$$

$$\therefore \; I_{in} = \frac{V_1}{R_1} + \frac{V_2}{R_2} + \frac{V_3}{R_3}$$

Just as with the inverting amplifier $I_{in} = I_f$

$$\therefore \; V_{out} = -(I_1 + I_2 + I_3) \times R_f$$

$$V_{out} = \left(V_1 \times \frac{R_f}{R_1} + V_2 \times \frac{R_f}{R_2} + V_3 \times \frac{R_f}{R_3} \right)$$

If all the resistors have the same value ($R_1 = R_2 = R_3 = R_f$) the output voltage will be the true sum of the applied inputs. If the resistors are not equal the output voltage will depend upon the ratio of each resistor and R_f. The circuit would then be described as a *weighted* summer or adder. This type of circuit is often employed as a digital-to-analogue converter.

The differential amplifier

This is where the op-amp provides an output signal that is proportional to the *difference* between the two input signals (Fig. 5.13).

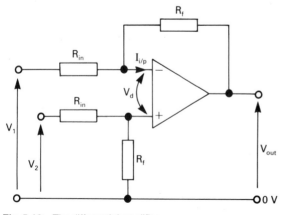

Fig. 5.13   The differential amplifier

Yet again our two rules apply and $I_{i/p} = 0$, $V_d = 0$ (virtual earth concept). Note that both input resistors (R_{in}) are the same and that the feedback resistor is the same value as the resistor joining the non-inverting terminal to the 0 V line. This has to be so in order that both the input terminals provide the same gain. A basic understanding of the op-amp as an amplifier tells us that

$$A_v = \frac{R_f}{R_{in}} \text{ and so } V_{out} = \frac{R_f}{R_{in}}(V_1 - V_2)$$

This can be easily verified by the next practical investigation.

Self Assessment 5

1   Calculate the values of R_{in} and R_f required to produce an inverting amplifier that has a voltage gain of 46 dB and an input resistance of 10 kΩ.
2   A non-inverting amplifier is required that has a voltage gain of 21 dB. Determine the values of R_{in} and R_f required if the input resistance is to be 10 kΩ.

Operational Amplifier Specifications

So far we have considered only the basic characteristics of the device. A cursory glance at the data sheets (pages 154 to 155) shows that there are a number of specifications quoted that require some explanation.

Common mode rejection ratio (CMRR)

The output of an op-amp is proportional to the difference between the voltages applied to the inverting and non-inverting input terminals. When these two voltages are the same, the output will be zero. A noise or interference voltage will be *common* to both terminals and should produce zero output. *BUT* due to the inputs having slightly different gains these *common mode* signals do not cancel entirely! The ability of an op-amp to suppress or reject these signals is its CMRR expressed in dB.

$$\text{CMRR} = 20 \times \text{Log}\left[\frac{\text{differential mode gain}}{\text{common mode gain}} \right] \text{dB}$$

The higher the CMRR value the better!

PRACTICAL INVESTIGATION *12*

The Op-Amp as a Differential Amplifier

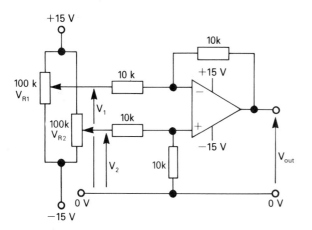

Equipment
Dual power supply
Three voltmeters
741 op-amp
4 × 10 k resistors
2 × 100 k potentiometers

Method
1 Connect the circuit as shown above.
2 Monitor V_1, V_2 and V_{out} with the voltmeters.
3 By varying V_{R1} and V_{R2} show that V_{out} is determined by V_1 and V_2.
4 Using a voltmeter measure V_{in} and V_{out} and show that $V_{out} = V_{in}$.
5 Adjust V_{R1} and V_{R2} so that $V_{out} = 0$ V and observe the result on V_{out} of
   (a)   making V_1 more positive than V_2
   (b)   making V_2 more positive than V_1.

Input offset voltage

Ideally when both inputs are zero the output voltage will be zero but due to internal imperfections in the device there will always be a small output voltage present — typically 5 mV. Most op-amps provide a facility for adjusting this *d.c. offset*, e.g. a variable resistor can be connected between pins 1 and 5 of the 741 (Fig. 5.14).

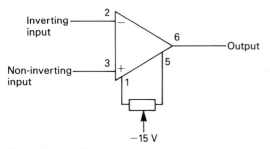

Fig. 5.14 d.c. offset

Input offset current

This is the minute current that is flowing in the input terminals even when the input voltages are zero (typically 30 nA).

Power supply rejection ratio (PSRR)

Changes in the power supply voltage will cause a change in the input offset voltage. The ratio of offset voltage change to the power supply voltage change is called the PSRR — usually quoted in dB and like CMRR the higher the number the better.

Slew Rate (*S*)

If an op-amp is fed with a step-input signal, i.e. a square wave, the output should change at the same rate as the input — *it does not do this!* The

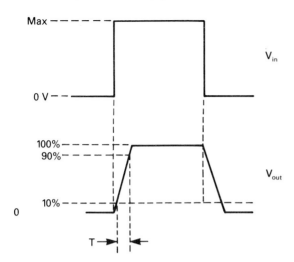

Fig. 5.15   The effects of 'slew rate'

result is a *slewing* of the output signal as shown in Fig. 5.15.

The slew rate (*S*) of an op-amp is expressed in volts per microsecond (V/μs) and indicates how quickly the device can respond to a change in input signal. This is usually measured and quoted as the time taken (*T*) for the output of a unity gain amplifier to change between two specified levels (10% and 90% of V_{out} max.) as indicated in Fig. 5.15. For example, supposing V_{out} max. was 10.0 V and the time (*T*) taken for the output to change from 1 V to 9 V was measured as 4 μs this would mean that V_{out} changed by 8 V in 4 μs.

$$\therefore \text{ the slew rate } S = \frac{8}{4} = 2 \text{ V/μs}$$

The slew rate for a 741 is specified as 0.5 V/μs while that of a LF351 is 13 V/μs. The greater the output change per microsecond that can be handled the better. Thus the LF351 has a superior slew rate to the 741.

This slew rate is of major significance and *not* only when step-inputs or square waves are being handled. The reason for this is that the response of an amplifier to a change in input will determine the frequency at which distortion will occur. Consequently it applies equally to sine wave signals since they have a peak-to-peak value that changes from positive to negative at a rate determined by the frequency. The slew rate will therefore determine

(a)   the maximum frequency before distortion occurs,
(b)   the maximum output signal that can be obtained without distortion.

It will thus dictate the bandwidth at which full power can be obtained from the op-amp.

Full power bandwidth

This specification is linked with the slew rate. It refers to the highest distortion-free frequency that can be obtained from an op-amp supplying the maximum output voltage sine wave. You will notice that the full power bandwidth (*FPB*) may be significantly less than the bandwidth calculated using the gain-bandwidth product.

$$FPB = \frac{S}{2\pi V_p}$$

Where S = Slew rate (in V/μs)
  V_p = peak output voltage
  FPB = full power bandwidth

Example
A 741 is to supply a distortion-free 20 V peak-to-peak sine wave signal.
Calculate the full power bandwidth.
Slew rate (*S*) for a 741 = 0.5 V/μs, V_p = 10 V

$$FPB = \frac{0.5/1 \times 10^{-6}}{2\pi \times 10 \text{ V}} = \frac{0.5 \times 10^6}{6.28 \times 10 \text{ V}}$$

$$= 7.962 \text{ kHz}$$

This means that distortion will occur at frequencies above 7.962 kHz. The full power bandwidth of a circuit can be increased by

(a)   reducing the output voltage (V_p),
(b)   using an op-amp with a better slew rate.

For example, if V_{out} only need be 10 V peak-to-peak

$$FPB = \frac{0.5 \times 10^6}{2\pi \times 5} = 15.9 \text{ kHz}$$

PRACTICAL INVESTIGATION **13**

Determination of Op-Amp Slew Rate

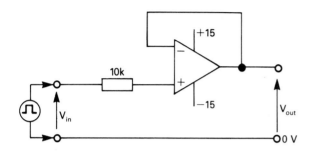

Equipment
Dual power supply
Dual beam oscilloscope
Signal generator
741 op-amp
LF351 op-amp

Method
1 Connect the circuit as shown above using the 741 op-amp.
2 Set the signal generator to give a 1 kHz square wave.
3 Monitor V_{in} and V_{out} and adjust the generator to give an easily measured V_{out} (e.g. 10 V).
4 Measure carefully and record the time taken for the output signal to change from 10% to 90% of its maximum value.
5 Calculate the slew rate (S) in V/μs.
6 Change to a sine wave and observe how V_{out} distorts as the frequency increases.
7 Repeat the above for the LF351 op-amp.

Results
1 Compare the practical slew rates of the 741 and the 351 op-amps with those quoted in the manufacturer's data sheet.

---

or if V_{out} is kept at 20 V peak-to-peak and LF351 is used then:

Slew rate of LF351 = 13 V/μs

$$FPB = \frac{13 \times 10^6}{2\pi \times 10} = 207 \text{ kHz}$$

Notice the difference?

THOUGHT _____

So when selecting an op-amp the higher the slew rate and full power bandwidth the better? — Not always, as with all devices the op-amp should be chosen to suit the application.

a.c. coupling

So far we have treated the op-amp as a direct coupled device (which internally is what it is). If it is desired to amplify only a.c. signals and block d.c. all that is required is to insert the appropriate value coupling capacitors. These will be in series with the input and output terminals as shown in Fig. 5.16.

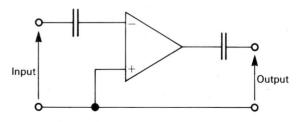

Fig. 5.16 Coupling capacitors

This will of course modify the frequency response curve and introduce a low frequency −3 dB point.

Selection of an Op-Amp

The data sheets (pages 154 and 155) show us there are many different types of device both bipolar and FET. Each offers different values of gain, bandwidth, slew-rate input and output resistances etc. The appropriate one to use will be determined by the application and the specifications that are required for a particular circuit. This can best be highlighted by examining a design case study.

Op-amp selection — A case study

It is required to design a circuit using a single op-amp to be used in the following situation.

The available signal is a 3 kHz 100 µV peak-to-peak sine wave that is supplied from a 10 MΩ source. The circuit designed to amplify this must provide a 50 mV peak output signal. The amplifier is to be battery operated from two 9 V dry cells so the power supply current for the op-amp must be as low as possible.

From this it appears there are two sets of specifications; those of the circuit supplying the signal and those relating to the circuit that is to be designed.

Signal specifications:
  Source resistance = 10 MΩ
  Signal frequency = 3 kHz
  Signal amplitude = 100 µV peak-to-peak
                (50 µV peak)
      Signal type = sine wave
      Bandwidth = 3 kHz (same as frequency)

Now for the circuit specifications:
  Input resistance = 10 MΩ (minimum)

Voltage gain A_v

$$A_v = \frac{V_{out}}{V_{in}} \qquad V_{out} = 50 \text{ mV peak}$$

$V_{in} = 50 \text{ µV peak}$

$$A_v = \frac{50 \times 10^{-3}}{50 \times 10^{-6}} = 1000 \text{ (60 dB)}$$

Bandwidth −3 kHz

CAREFUL! − The gain A_v must be *60 dB* (1000) up to 3 kHz, so the chosen op-amp must have an open loop gain much larger than this! The op-amp bandwidth can be calculated using the gain-bandwidth product

GB = gain × bandwidth (this is a constant!)

gain = 1000   bandwidth = 3 kHz

$GB \times 3 \times 10 = 3 \text{ MHz}$.

Now this shows that the gain-bandwidth product is 3 MHz. So the chosen op-amp must have unity gain at 3 MHz.

$\therefore f_T = 3 \text{ MHz}$ (min)   or   the   unity   gain
                bandwidth = 3 MHz (min)

(remember f_T = unity gain bandwidth!)

Slew rate
This is not going to to be very significant since V_{out} is only 50 mV peak, but let us go through the motions using the 3 kHz bandwidth even though our circuit will not be operating at full power.

from $FPB = \dfrac{S}{2\pi V_P}$

$$S = 2 \, FPB \, V_P$$
$$FPB = 3 \text{ kHz}$$
$$V_P = 50 \text{ mV}$$

$\therefore \ S = 6.28 \times 3 \times 10^3 \times 50 \times 10^{-3}$
        $= 942 \text{ µV/µs} \simeq 1 \text{ mV/µs}$.

The slew rate of all op-amps is better than this!

Power supply current

The circuit is to be supplied from batteries so the lowest supply current possible is to be aimed for here. The choice will be between FET and bipolar devices but will not simply be a matter of choosing the device with the minimum supply current, this must be considered against the other required specifications.

We are now in a position to state the desired device specifications and make a choice.

Required specifications

Input resistance = 10 MΩ minimum
Voltage gain (A_v) = high, e.g. 100 dB (60 dB is the gain we require with negative feedback applied)
Power rupply range = ±9 V
Slew rate = 1 mV/μs
Unity gain bandwidth
(f_T) = 3 MHz (min)
Supply current = minimum, i.e. less than 3 mA.

How to choose

The major limitations must be considered initially. In our case they are the input resistance at 10 MΩ and the F_T at 3 MHz (min). Look for a device that fulfils these specifications first. On the basis of these two it appears that an FET device will have to be chosen.

Now consider a device that fulfils the gain, slew rate and power supply requirement. It appears that an LF351 or similar device will be suitable.

LF351
Input resistance   − 1 TΩ (1×10^{12} Ω)
Voltage gain (A_v)   − 100 dB
Power supply range   − ±5 to ±18 V
Slew rate   − 13 V/μs
Unity gain bandwidth − 4 MHz
Supply current   − 1.8 mA

All that now remains is to design the circuit around the chosen device using the techniques mentioned earlier.

Operational Amplifier Review

1 The op-amp is an integrated circuit differential amplifier.
2 It has a very high open loop voltage gain (A_v = 100 dB).
3 Direct coupling gives a low frequency cut-off of d.c. (0 Hz).
4 The upper frequency limit is specified as the frequency at which the gain is unity − quoted as the *unity gain bandwidth* or the *transition frequency* (f_T).

5 A positive and negative power supply is usually required; typically ±15 V.
6 To use as a practical amplifier the op-amp has negative feedback applied, thus reducing the gain and widening the bandwidth.
7 The closed loop bandwidth and gain can be determined using the gain-bandwidth product ($GB = A_v$ numerical × bandwidth).
8 Due to its very high open loop gain the closed loop gain of an op-amp circuit is independent of gain (A_v) and is determined by the ratio of R_f and R_{in}.
9 The gain (A_v) of an inverting amplifier is determined using

$$A_v = -\frac{R_f}{R_{in}}.$$

10 The gain (A_v) of a non-inverting amplifier is determined using

$$A_v = \left(1 + \frac{R_f}{R_{in}}\right)$$

11 A unity gain *buffer* amp can be built using an op-amp and *no* external components.
12 An op-amp can be a.c. coupled by using series capacitors to block d.c.
13 The slew rate (S) is quoted in V/μs and indicates the speed of response of the device to a change in input signal. $S = 2\pi FPB\, V_p$.
14 The full power bandwidth is the highest output frequency signal that can be supplied without distortion and is related to the slew rate

$$FPB = \frac{S}{2\pi V_p}.$$

15 Op-amps like other components are available in a vast range with each device offering different characteristics.
16 An op-amp should be chosen to suit the required applications and where possible it is wise to use the common type of device, i.e. 741, to minimize the cost.

Self Assessment Answers

Self Assessment 5

1 $R_{in} = 10\,k\Omega$ (input resistance)

$$A_v = \frac{R_f}{R_{in}} \qquad A_v = 46\,dB$$

numerical gain $= A.\log\left(\dfrac{46}{20}\right) = 199.5 \approx 200$

$R_f = R_{in} \times A_v = 10 \times 10^3 \times 200 = 2\,M\Omega$
nearest preferred value (npv) 2M2.

2 $R_{in} = 10\,k\Omega$ (to give an input resistance of 10 kΩ)

$$A_v = 1 + \frac{R_f}{R_{in}} \quad \therefore \frac{R_f}{R_{in}} = A_v - 1$$

$A_v = 21\,dB$

numerical gain $= A.\log\left(\dfrac{21}{20}\right) = 11.2 \approx 11$

$\dfrac{R_f}{R_{in}} = 11 - 1 = 10$

$R_f = R_{in} \times 10 = 10\,k\Omega \times 10 = 100\,k\Omega$

6

Noise

In an electronic system noise *is defined as any unwanted signal that is present along with the wanted signal.*

The presence of noise worsens the performance of any system — *crackles* and *pops* can spoil our enjoyment of a radio broadcast as can *clicks and hiss* from a record or *hum* from an amplifier. Unlike distortion, however, noise is present even when there is no signal. The terrible truth is that noise will *always* be present in a system, and cannot be totally removed. Measures can be taken, however, to ensure that noise is kept to a low level and so will offer minimal disturbance to the operation of a system. This results in a noise level that is low enough to be masked by the actual signal level.

Noise can be classified into two types: external and internal noise as indicated by Fig. 6.1.

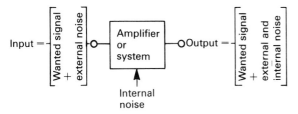

Fig. 6.1   Noise in an electronic system

External Noise

This consists of unwanted signals which enter the system to be processed along with the wanted signals. The sources of external noise are quite numerous some of which are completely natural, like static or atmospheric disturbance often experienced on radio and TV as caused by thunderstorms, or the cosmic electromagnetic radiation that is produced by the sun and the stars. Most causes of external noise, however, are man-made and we shall now examine these in detail, together with methods that can be used to reduce their effects.

Mains 'Hum'

Any mains operated electronic system will probably incorporate a power supply that produces a suitable d.c. operating voltage. The a.c. mains has a frequency of 50 Hz and this 50 Hz signal can be *picked up* before and after rectification to appear at the output of the system as an annoying 50 Hz or 100 Hz *hum*.

Precautions against mains hum

Mains *pick-up* is due mostly to electromagnetic induction occurring between the mains wires and the circuit wiring. To reduce this the following steps can be taken:

- Ensure that the mains leads are positioned as far away as possible from the low voltage side of the circuit.
- Use screened leads for the input leads to the system.
- The electronic circuit can be physically *shielded* from the mains side by the use of a metal cover or plate that is earthed.
- All earths should be taken to a common point.
- Low voltage d.c. supply lines can be decoupled to earth using capacitors that have a very low reactance at mains frequency and so will remove any 50 Hz or 100 Hz signal, that may be present (Fig. 6.2).

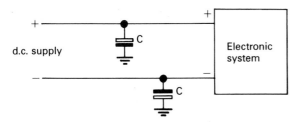

Fig. 6.2 Decoupling a.c. signals from d.c. lines

Switching or contact noise

Arcs or sparks, however caused, are major sources of noise. If you doubt this try the following:

Press a piezo-electric lighter several times near an operating television receiver and observe the lines that appear on the screen as the lighter *clicks*. Alternatively, while the television is on, operate a mains light switch several times and note the interference caused.

This shows that whenever a flow of current is interrupted a burst of electromagnetic radiation occurs that contains multiple frequencies up to hundreds of megahertz. Should these frequencies enter a system, interference will result. Types of equipment that are likely to generate switching noise include:

- Electric motor — (sparking brushes)
- Car ignition system — (contact breakers)
- Fluorescent lights — (starting mechanism)
- Solenoid valves on boiler and control equipment — (high switching currents as valve operates)
- Medical diathermic equipment — (high switching currents).

Precautions

If the noise is generated by a switching operation then one solution is for the equipment producing the noise to have suppressors wired across the switch contacts or brushes. This is often just a simple capacitor-inductor filter that prevents the high frequencies radiating. It is now law that all equipment sold which is likely to produce this type of interference must include suppression to minimize radiated interference. Some apparatus, e.g. arc welding equipment, cannot be suppressed however. In addition to this there will always be equipment operating that has insufficient or faulty suppression. It is wise to protect sensitive equipment from this external interference by using screened leads for all the input cables and shielding or screening the low voltage electronic circuits.

THOUGHT

So the general rule seems to be if in doubt screen thoroughly? — Generally this makes sense, screening is like a cage, it can stop things getting out as well as getting in! Where a lot of arc welding is carried out in a permanent location an earthed screen is incorporated in the plastered walls of the workshop, this prevents interference radiating out. Likewise sensitive equipment is often placed in a *cage like* structure to prevent interference getting in (Fig. 6.3).

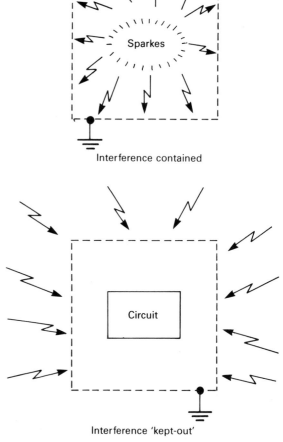

Interference contained

Interference 'kept-out'

Fig. 6.3 Screening for protection against radiated interference

This type of screening cage can consist of wire mesh (similar to chicken wire) construction. This is called a *Faraday cage* and curiously enough low frequency radiated interference views the mesh as a solid wall and cannot pass through.

To screen a circuit from frequencies up to about 100 kHz, 1 cm mesh is adequate, above this frequency the radiation is more penetrative, with progressively smaller mesh size required until VHF frequencies are reached when sheet material screening is required, i.e. foil or plate. At UHF the penetration of any interference is greater so holes and cracks or joints in the screened box are not permitted. Special lead through capacitors have to be used for connections to the screened circuit. Above UHF, screening becomes much more involved since electromagnetic radiation at these frequencies can penetrate and pass through normal screening materials, so double wall screening may be required at super and extra high frequencies and only lead will protect against X-rays and above.

Crosstalk

It is often convenient to send information using a pair of conductors as in telephone and data transmission systems. For economical reasons multicore cables are used with a number of pairs sharing a common cable. *Crosstalk* is the signal pick-up that occurs between pairs in the cable.

Precautions against crosstalk

To reduce this type of interference several methods are used. Pairs that are individually screened within the common cable, multiway ribbon cables that alternate earth conductors with the signal conductors in the ribbon and multiway cables that use twisted pair construction. In addition to these methods it is usual to reduce the cross-coupling between pairs by keeping signal amplitudes low.

Natural noise

This was mentioned previously and consists of the static generated by atmospheric conditions, i.e. electrical and magnetic storms. There is also galactic noise from the stars and a certain amount of earth noise is radiated by the planet on which we live.

Precaution

These noises are ever present and are only likely to cause interference on systems that use aerials. Minimization of the effects of noise of this type consists of using screened or coaxial cables together with careful aerial design and positioning.

Internal noise

This is noise that is introduced by the system itself! It is worth mentioning from the start that *all* electronic circuits will introduce noise since it is generated by the very components and wires themselves. A *noise free* electronic circuit is a myth and cannot ever exist in practice. The best that can be achieved is a circuit that generates a very low level of *noise* that is not detected in the output. Now is the time to consider the various causes of internal noise.

Thermal noise

Also called *Johnson noise*, this is the noise voltage caused by the random motion of electrons in electrical conductors and components. Electrons form part of the atomic structure of conductors. At absolute zero ($-273°C$) the atoms can be considered to be in a state of rest but as the temperature rises above this, thermal agitation occurs and electrons tear free from the atoms' outer orbits. These *free electrons* are now at liberty to wander randomly throughout the atomic structure. Since any electron movement constitutes a flow of current an EMF or voltage will be set up across the conductor. This voltage is thermally generated and so will increase with a rise in temperature. At temperatures above absolute zero (0 K), considerable random electron movement will be taking place causing a small but significant noise voltage to be present in all components and conductors. The rms value of this noise voltage can be calculated using Johnson's equation

$$V_n = \sqrt{4\,kTBR} \quad \text{Volts.}$$

Where:

V_n = rms value of the noise voltage.
 k = Boltzmans constant 1.38×10^{-23} Joules/Kelvin.
 T = Absolute temperature in kelvin (K = °C + 273)
 B = Bandwidth in hertz.
 R = Resistance in ohms.

You can see from the equation that the amount of noise produced in a component is determined not only by the temperature, but by the bandwidth, i.e. the frequency (or range of frequencies) over which it operates and the components' resistance.

Example 1

A $10\,k\Omega$ resistor at room temperature (20°C) operates at 1 MHz. Calculate the thermal noise generated.

$V_n = \sqrt{4\,kTBR}$
 $k = 1.38 \times 10^{-23}$
 $T = 20°C + 273 = 293\,K$
 $B = 1\,MHz$
 $R = 10\,k$
 $V_n = \sqrt{4 \times 1.38 \times 10^{-23} \times 293 \times 1 \times 10^6 \times 10 \times 10^3}$
 $\quad = 12.7 \times 10^{-6}\,V.$

$\therefore V_n = 12.7\,\mu V$ (rms)

Now consider the same resistors operating at 2 MHz.

$V_n = 17.98\,\mu V$ (rms)

It becomes clear from this that an amplifier with a wide bandwidth will generate more thermal noise than one with a narrow bandwidth.

Example 2

An electrical component has a resistance of $200\,k\Omega$. If its operating frequency is 5 MHz, calculate the thermal noise generated at, (a) 20°C (b) 50°C.

(a) $V_n = \sqrt{4\,kTBR}$
 $\quad = \sqrt{4 \times 1.38 \times 10^{-23} \times 293 \times 5 \times 10^6}$
 $\qquad\qquad\qquad\qquad\qquad \times 200 \times 10^3$

 $V_n = 127.1\,\mu V$

(b) $V_n = \sqrt{4 \times 1.31 \times 10^{-23} \times 323 \times 5 \times 10^6}$
 $\qquad\qquad\qquad\qquad\qquad \times 200 \times 10^3$

 $V_n = 133.5\,\mu V$

From this, the need to keep the temperature of circuits as low as possible is apparent, particularly when you realize that the temperature is governed not only by ambient conditions but also by the current flowing in a circuit.

Thermal noise powers

We are ultimately interested in the actual noise power that is transferred from a source into a load. This is best achieved by considering any source or input to a system as a noise generator feeding the system (Fig. 6.4).

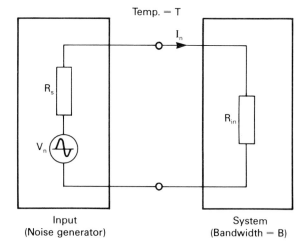

Fig. 6.4   Noise power diagram

Maximum power transfer occurs when source and load are matched as in this case

$R_s = R_{in} = R$

This will also be the condition for maximum noise power.

Noise current $I_n = \dfrac{V_n}{R_s + R_{in}}$

Since $R_s = R_{in} = R$

$I_n = \dfrac{V_n}{2R}$

If the noise power transferred into the input of the system $= P_n$

then:

$$P_n = I_n^2 \times R_{in}$$

$$\therefore P_n = \left(\frac{V_n}{2R}\right)^2 \times R_{in} \qquad (R_{in} = R)$$

$$\therefore P_n = \frac{V_n^2}{(2R)^2} \times R \qquad \text{So } P_n = \frac{V_n^2}{4R}$$

Now

$$V_n = \sqrt{4\,kTBR} \quad \therefore V_n^2 = 4\,kTBR$$

$$So \qquad P_n = \frac{\cancel{4}\,kTB\cancel{R}}{\cancel{4}\cancel{R}}$$

So $P_n = kTB$ watts

THOUGHT

So when a source is matched to its load the noise power is independent of resistance — Quite so! This is a very useful concept that can be used to determine the maximum thermal noise power that will be transferred when source and load are matched.

Shot noise

This is noise that occurs in transistors and active devices due to the arrival and departure of charge carriers at junctions within the device. The rms noise current (I_n) can be calculated using:

$$I_n = 2\,eIB \text{ Amps}$$

Where

e = charge on an electron 1.6×10^{-19} C
I = d.c. current (A)
B = Bandwidth (Hz)

Example

Calculate the noise current produced by a diode operating at 15 kHz when passing a current of (a) 250 mA (b) 500 mA

(a) $I_n = 2 \times 1.6 \times 10^{-19} \times 250 \times 10^{-3} \times 15 \times 10^3$
   $= 1.2 \times 10^{-15}$ A

(b) $I_n = 2 \times 1.6 \times 10^{-19} \times 500 \times 10^{-3} \times 15 \times 10^3$
   $= 2.4 \times 10^{-15}$ A.

This indicates that I_n is directly proportional to the d.c. current and so the amount of shot noise can be determined by the device's operating current.

Although shot noise can be a source of major irritation for sensitive electronic circuits there is sometimes a need for a noisy component. When building a specific sound generator like a music synthesizer, a noise diode would be used for the production of drum and cymbal sounds.

Flicker noise

This is a low frequency noise that occurs in semiconductors due to the random way electron-hole pairs are generated and recombined. Flicker noise power is inversely proportional to frequency with power levels virtually insignificant above about 15 kHz. This noise is most apparent in audio frequency systems.

Partition noise

This again occurs in semiconductors. It is produced due to fluctuations in the way current divides at the junctions, e.g. in a bipolar junction transistor $I_E = I_C + I_B$, the emitter current (I_E) divides to provide the base and collector current I_B and I_C. This division of current varies giving rise to partition noise.

Precautions against internal noise

From this brief study of the causes of internal noise it can be seen that there is very little the designer can do to guard against it. Since it is generated by the components themselves, internal noise can best be minimized by selecting special *purpose-made* low-noise components for sensitive circuits and ensuring that a low operating temperature is maintained.

Noise Colours

If you have read much on electronics you may have been puzzled by reference to *white* or *pink* noise. It has become widely accepted to use these colours to categorize the types of noise that can be encountered. This is analogous to the visible light spectrum. White light comprises all the colours of the spectrum and white noise contains noise of all frequencies. Red light is light of the lowest visible frequency and so the expression pink noise refers to low-frequency noise. Let us now consider these two definitions in greater detail.

White noise

Noise is random, and if it has a *flat* or level power/frequency spectrum it is considered to be white noise (Fig. 6.5).

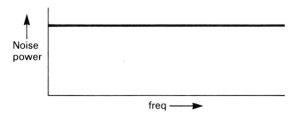

Fig. 6.5 'White' noise

This means that at any frequency over a given range, there is the same amount of noise power present. Thermal noise, shot noise and partition noise are all examples of white noise.

Pink noise

This is noise with a power that is inversely proportional to frequency. Sometimes called $1/f$ noise. Its power/frequency spectrum is as shown in Fig. 6.6.

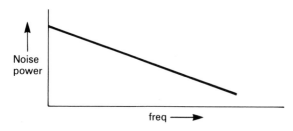

Fig. 6.6 'Pink' noise

As the frequency increases the noise power reduces. Flicker noise is an example of pink noise although it is perfectly possible to filter white noise in order to produce pink noise.

The Signal-to-Noise Ratio

Noise is always present and will exist at the input to a system. It will also be produced by the system itself and so appear in greater quantities at the system output. The significance of any noise power can only be judged by comparing it to the wanted signal power. This is achieved using the signal-to-noise ratio.

$$S/N \text{ ratio} = \frac{\text{wanted signal power}}{\text{unwanted noise power}}$$

$$= \frac{\text{Signal power}}{\text{Noise power}}$$

Example

The input signal to an amplifier is 120 mW with 5 mW of noise present. Calculate the *S/N* ratio.

$$S/N = \frac{120 \times 10^{-3}}{5 \times 10^{-3}} = 24{:}1$$

This indicates that the signal level is 24 times greater than the noise level. Obviously the greater the *S/N* ratio the better because then the unwanted noise will be much lower than the wanted signal level.

THOUGHT _____

Consider the situation that would result from a *S/N* ratio of 1:1. This would mean that both the signal and noise powers would be the same and there would be no way of sorting the signal from the noise — situation hopeless!

Signal-to-noise ratio in decibels

It is very convenient to express this ratio in terms of decibels (dB) in exactly the same way as gain is expressed. The signal-to-noise ratio is then given by:

$$S/N = 10 \, Log \left(\frac{\text{Signal power}}{\text{Noise power}} \right) dB$$

$$= 10 \, Log \left(\frac{S}{N} \right) dB$$

Expressing the given example in this way gives:

$$S/N \text{ ratio} = 10 \, Log \left(\frac{120 \times 10^{-3}}{5 \times 10^{-3}} \right) dB$$

$$S/N = 13.8 \, dB$$

The significance of the *S/N* ratio depends upon the requirements of a given application. For example, *S/N* ratios of about 12 dB are acceptable for the telephone system where the message can be readily understood despite the presence of considerable noise levels. A good-quality hi-fi system would require a signal-to-noise ratio of about 60 dB or greater in order to provide a high-quality music output — consult your systems handbook and see how it measures up!

THOUGHT

So the higher the S/N ratio the better?
Correct. 20 dB is better than 15 dB and 70 dB is better than 60 dB.

Self Assessment 6

1 (a) The input signal power to an amplifier is 120 mW with 60 μW of noise present. Calculate the input signal-to-noise ratio in dB.
   (b) The same amplifier has an output *S/N* ratio of 28 dB. If the output signal power is 2.5 W calculate the output noise power.
2 The noise voltage at the output of an amplifier is 82 mV while the signal voltage is 1.2 V. Calculate the output *S/N* ratio in dB.

Input and output signal-to-noise ratios

By referring to the section on the sources of noise it is hopefully clear that any system will produce its own internal noise, so the output *S/N* ratio

will always be worse (lower) than the input S/N ratio.

It is important to quantify the internally produced noise since the noise at the output of a system will consist of the processed input noise and the internal noise. The internal noise of a system can be dealt with in two ways:

1 The internal noise can be treated as a separate entity that is produced by the system as indicated by Fig. 6.7.

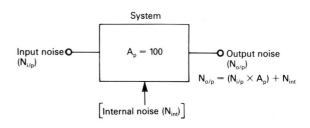

Fig. 6.7   Internal noise in a system

You can see that using this method the internal noise is simply added to the input noise *after* it has been processed by the system, the internal noise is *NOT* itself processed.

2 The second method treats the internal noise as though it were a separate input noise (Fig. 6.8).

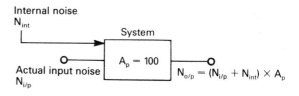

Fig. 6.8   Internal noise referenced to the input of the system

In this way the noise that is internally generated is referenced to the input terminals of the system and so is processed in the same way as the actual input noise. These two methods are confusing, so let's use an example to help clarify the situation.

Example 1

An amplifier has an input signal of 100 mW with 50 µW of noise present. If the amplifier has a gain $A_p = 20$ dB and introduces 2 mW of internal noise, determine the output noise power.

$$N_{i/p} = 50 \text{ µW}$$
$$A_p = 20 \text{ dB} = 100 \text{ (numerical gain)}$$
$$N_{int} = 2 \text{ mW}.$$
$$\therefore \quad N_{o/p} = (N_{i/p} \times A_P) + N_{int}$$
$$= (50 \times 10^{-6} \times 100) + 2 \times 10^{-3} = 7 \text{ mW}$$

$\therefore$ Output noise power $= 7$ mW.

Now consider the same situation but worded slightly differently.

Example 2

An amplifier with a power gain of 20 dB has an input signal of 100 mW with 50 µW of noise present. If the amplifier itself generates additional noise power of 20 µW referred to its input terminals, calculate the output noise power.

Now

$N_{i/p} = 50$ µW $A_p = 20$ dB $= 100$

$N_{int} = 20$ µW *but* this is treated as an additional input noise

$$N_{o/p} = (N_{i/p} + N_{int}) \times A_p$$

$$= (50 \text{ µW} + 20 \text{ µW}) \times 100 = 70 \times 10^{-6} \times 100$$
$$= 7 \text{ mW}$$

Output noise power $= 7$ mW

So you see the answer is the same but there is a subtle difference in how it is calculated. Be sure you *understand* that difference!

Noise Figure (*F*)

Sometimes called the noise factor; this is used to define how much noise a system introduces, i.e. it is an indication of how a system *worsens* the input signal-to-noise ratio.

Noise factor (*F*)

$$= \frac{\text{total output noise power}}{\text{output noise power due to input noise power}}$$

Ideally the noise factor of any system would be unity, indicating that the system introduces no noise at all, but we know this to be impossible!

A useful way of determining the noise factor is to treat it as the ratio between the input and output signal-to-noise ratio.

$$F = \frac{\text{Input } S/N \text{ ratio}}{\text{Output } S/N \text{ ratio}}$$

Now because it is usual to express *S/N* ratios in decibels which are themselves logarithmic ratios, the division becomes a simple subtraction such that

Noise factor (*F*) = Input *S/N* ratio in dB − output *S/N* ratio in dB

Example

An amplifier has an input *S/N* ratio of 60 dB and an output *S/N* ratio of 48 dB. Calculate the noise factor of the amplifier.

$$F = 60 \text{ dB} - 48 \text{ dB} = 12 \text{ dB}.$$

You now have all the information needed to perform quite involved problems concerning signal-to-noise ratios.

Example

The input signal to a 30 dB amplifier is 20 µW. The amplifier itself produces 8 nW of internal noise referred to its input. If the input signal-to-noise ratio is 40 dB calculate:

(a)   The input noise power to the amplifier.
(b)   The total output noise power.
(c)   The output signal power.
(d)   The output signal-to-noise ratio.
(e)   The noise factor of the amplifier.

(a)   The input noise power
      Input signal power $= S_{i/p} = 20$ µW
      Input *S/N* ratio $= 40$ dB
      Step 1   Convert *S/N* ratio to a numerical value

$$S/N \text{ ratio} = A \cdot \log\left(\frac{40}{10}\right) = 10 \times 10^3$$

Step 2 Calculate the noise power by transposing the formula

$$\frac{S_{i/p}}{N_{i/p}} = 10 \times 10^3$$

$$N_{i/p} = \frac{S_{i/p}}{10 \times 10^3} = \frac{20 \times 10^{-6}}{10 \times 10^3} = 2 \text{ nW}$$

Input noise power = 2 nW
(b) The total output noise power
Step 1 Decide which equation to use

$$N_{o/p} = (N_{i/p} + N_{int}) \times A_p$$
$$N_{int} = 8 \text{ nW (referred to input terminals)}$$

Step 2 Convert A_p to a numerical value

$$A_p = 30 \text{ dB} \quad A . \text{Log}\left(\frac{30}{10}\right) = 1000$$

Step 3 Calculate $N_{o/p}$

$$N_{o/p} = (2 \text{ nW} + 8 \text{ nW}) \times 1000 = 10 \text{ } \mu\text{W}$$

Output noise power = 10 µW
(c) The output signal power
This is simply the input signal power × A_p

$$S_{i/p} = 20 \text{ } \mu\text{W} \quad A_p = 1000$$
$$S_{o/p} = 20 \text{ } \mu\text{W} \times 1000 = 20 \text{ mW}$$
$$\therefore \text{Output signal power} = 20 \text{ mW}$$

(d) Output signal-to-noise ratio

This is given by $10 \text{ Log}\left(\dfrac{S_{o/p}}{N_{o/p}}\right)$ dB

$$= 10 \text{ Log}\left(\frac{20 \times 10^{-3}}{10 \times 10^{-6}}\right) \text{ dB} = 33 \text{ dB}$$

$\therefore$ Output signal-to-noise ratio = 33 dB
(e) Noise factor
F = Input *S/N* ratio in dB − Output *S/N* ratio in dB
F = 40 dB − 33 dB = 7 dB
Noise factor = 7 dB

Self Assessment 7

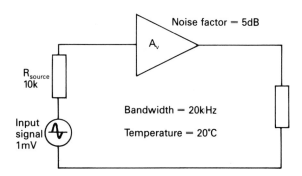

Noise factor = 5dB

A_v

R_{source} 10k

Bandwidth = 20kHz

Input signal 1mV

Temperature = 20°C

From the diagram shown calculate:
1 The input noise power.
2 The input signal-to-noise ratio.
3 The output signal-to-noise ratio.

Noise Review

1 Noise is defined as any unwanted signal.
2 Noise can be external or internal to a system.
3 External noise is produced by phenomena outside the system and categorized as *natural* or *man-made*.
4 Sources of *natural* noise include: static, thunderstorms, atmospheric disturbance, cosmic and galactic radiation.
5 Sources of *man-made* noise include mains *hum*, contact or switching noise, crosstalk.
6 Precautions against external noise include using screened leads, shielding or screening sensitive circuits and suppression of equipment likely to cause interference.
7 External noise is generated by the components and wires of the system itself. It can be defined using the following categories: Thermal or Johnson noise, shot noise, flicker noise and partition noise.
8 Thermal or Johnson noise voltage can be calculated using

$$V_n(\text{rms}) = \sqrt{4 \, kTBR}.$$

9 Shot noise current can be calculated using

$$I_n(\text{rms}) = 2 \, eIB$$

10   Where a source is matched to its load the maximum thermal noise power that can be transferred is given by

$$N_P = kTB \text{ Watts}$$

11   Precautions against internal noise are limited to careful design, using high-quality components and maintaining low operating temperatures.

12   Noise can be categorized using colours.

13   White noise has a level power frequency spectrum, i.e. at any frequency the noise power is the same. Thermal, shot and partition noise is white.

14   Pink noise varies inversely with frequency ($1/f$). Flicker noise is pink.

15   The signal-to-noise ratio indicates the strength of the wanted signal when compared to the strength of the unwanted noise present. Usually expressed in dB using

$$S/N \text{ ratio} = 10 \text{ Log} \left(\frac{\text{Signal power}}{\text{Noise power}} \right) \text{ dB}$$

or $SN \text{ ratio} = 20 \text{ Log} \left(\frac{\text{Signal voltage}}{\text{Noise voltage}} \right) \text{ dB}$

16   The higher the *S/N* ratio the *cleaner* will be the signal.

17   The output *S/N* ratio of any system will *always* be lower (worse) than the input *S/N* ratio due to the internal noise produced.

18   The noise figure or factor indicates how much a system will degrade the input *S/N* ratio

$$F = \frac{\text{Input } S/N \text{ ratio}}{\text{Output } S/N \text{ ratio}} \quad \text{or in dB,}$$

$$F = (\text{Input } S/N \text{ ratio} - \text{Output } S/N \text{ ratio})$$

Self Assessment Answers

Self Assessment 6

1   (a) $S/N \text{ ratio} = 10 \text{ Log} \left(\frac{\text{Signal power}}{\text{Noise power}} \right) \text{ dB}$

$$= 10 \text{ Log} \left(\frac{120 \times 10^{-3}}{60 \times 10^{-6}} \right) \text{ dB} = 33 \text{ dB}$$

(b) $S/N \text{ ratio} = 28 \text{ dB} \quad 28 \text{ dB} = 10 \text{ Log} \left(\frac{S}{N} \right)$

so $A. \text{Log} \left(\frac{28}{10} \right) = \frac{S}{N} = 631 \text{ (as a ratio)}$

$$\frac{S}{N} = 631 \qquad \text{Noise power} = \frac{\text{Signal power}}{631}$$

Signal power $= 2.5 \text{ W} \quad \text{Noise} = \frac{2.5}{631} = 3.96 \text{ mW}$

Output noise power $= 3.96 \text{ mW}$

2   *Note* in this question the quoted levels are voltages rather than powers!
Signal level $= 1.2 \text{ V}$
Noise level $= 82 \text{ mV}$

$$S/N \text{ ratio} = 20 \text{ Log} \left(\frac{\text{Signal voltage}}{\text{Noise voltage}} \right) \text{ dB}$$

$$= 20 \text{ Log} \left(\frac{1.2}{82 \times 10^{-3}} \right) \text{ dB} = 23.3 \text{ dB}$$

An output *S/N* ratio $= 23.3 \text{ dB}$

Self Assessment 7

1   $N_{i/p} = $ thermal noise voltage given by
$V_n = \sqrt{4 kTBR}.$
$= \sqrt{4 \times 1.38 \times 10^{-23} \times 293 \times 20 \times 10^3 \times 10 \times 10^3}$
$= 1.8 \text{ μV}$
Input noise power $= 1.8 \text{ μV}$

2   Input signal-to-noise ratio

$$= 20 \text{ Log} \left(\frac{S_{i/p}}{N_{i/p}} \right) = 20 \text{ Log} \left(\frac{1 \times 10^{-3}}{1.8 \times 10^{-6}} \right)$$

$$= 55 \text{ dB}$$

Input *S/N* ratio $= 55 \text{ dB}$

3   Output signal-to-noise ratio
$F = (\text{Input } S/N \text{ ratio} - \text{Output } S/N \text{ ratio}) \quad (\text{decibels})$
Output *S/N* ratio $= (\text{Input } S/N \text{ ratio} - F)$
$F = 5 \text{ dB}$
Output *S/N* ratio $= (55 \text{ dB} - 5 \text{ dB}) = 50 \text{ dB}$
Output *S/N* ratio $= 50 \text{ dB}.$

7

Oscillators

An oscillator is an electronic circuit that produces an output signal without any external input signal.

The output from an oscillator may be sinusoidal or non-sinusoidal waveforms that are rectangular, square, triangular or saw tooth. We will concentrate on sine wave oscillators.

The Oscillator Principle

The section on amplifiers and feedback stresses the importance of stability in an amplifier and highlights the fact that to make an amplifier stable, negative feedback is always used. Further to this is the fact that should positive feedback occur in an amplifier circuit there is a strong chance it will become unstable at a certain frequency and start to oscillate! To create an oscillator, simply take an amplifier and give it sufficient positive feedback to make it become unstable and produce oscillations. Consider the diagram in Fig. 7.1.

Let us recap on the feedback theory involved.

$A_{pfb} = \dfrac{A_v}{1 - \beta A_v}$ this is the positive feedback equation.

Now if $\beta A_v = 1$ the equation becomes

$A_{pfb} = \dfrac{A_v}{1 - 1} = \dfrac{A_v}{0} = \infty$

this is the unstable state and oscillations will occur.

The output frequency

Oscillators are usually designed to produce an output at a specific frequency (this is sometimes variable over a given range). From Fig. 7.1 it can be seen that the amplifier derives its input signal from its own output. It is vital that the feedback network (β) allows only the required oscillation frequency to be fed back to the input. If a 1 kHz output is required the only permitted feedback signal (βV_{out}) must be at 1 kHz. Consequently the circuit that provides the positive feedback must also determine the frequency. Once the required feedback has been achieved it must be remembered that the amplifier's input signal is supplied from its own output and should this be reduced then the input will also fall and oscillations will *peter out* and stop. The circuit must therefore provide sufficient voltage gain to maintain its own oscillations.

From the foregoing information it becomes obvious that there are some essential requirements that an oscillator circuit must possess if it is to be successful. These are as follows:

1 A d.c. power supply.
2 Positive feedback (pfb) of 0° or 360° within the closed loop.
3 A frequency determining network — circuitry to ensure that pfb occurs at only one

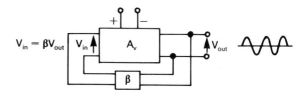

Fig. 7.1 Block diagram of an oscillator

frequency and at this frequency $\beta A_v = 1$.
4 Sufficient gain to maintain oscillations.

If these requirements are fulfilled, your oscillator will work!

THOUGHT

This is fine but how does the oscillator start when the power is switched on? After all at switch on there is no input signal so there will be no output and no feedback signal!
Initially this appears to be the case but if you refer to the section on noise it is apparent that at the input to the amplifier there will be a tiny amount of *white noise* present (i.e. a noise voltage containing all frequencies). This input noise will be amplified and so will appear larger at the output. A portion of this noise voltage will be fed back to the input via the frequency determining network. The required frequency will appear as a very small input voltage; this in turn will be amplified appearing larger at the output, to be fed back, amplified etc. etc., and so the circuit will *burst* into life as illustrated in Fig. 7.2.

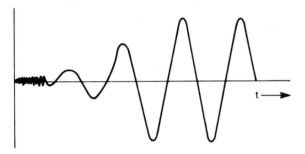

Fig. 7.2   Oscillator starting waveform

Positive feedback

How this is achieved in practice will be determined by the amount of phase shift an amplifier produces between input and output. If the amplifier is non-inverting then it introduces no phase shift. It is then simply a matter of feeding back some of the output signal (βV_{out}) in order to produce positive feedback.

If, however, an inverting amplifier is used it will introduce 180° of phase shift. Should a fraction of this be fed back it would be in anti-phase and so negative feedback would result. In a case such as this a further 180° of phase shift must be introduced in order to produce

positive feedback, i.e. 180° amplifier phase shift + 180° phase shift = 360° = 0° = positive feedback.

There are some amplifier circuits that will introduce a phase shift between 0° and 180°. Obviously to provide positive feedback in such situations will involve ensuring that the total phase shift from output back to input is zero degrees. We shall concentrate only on oscillators that use inverting and non-inverting amplifiers.

The frequency determining network

This is a critical part of the oscillator circuit since it determines the frequency of the output signal. Typically, such networks incorporate tuned resonant LC circuits, or a phase shift arrangement that uses resistors and capacitors. To give some credence to the theory, let us now consider some practical circuits.

The Tuned Collector Oscillator

Look carefully at Fig. 7.3 and you will be able to see the familiar tuned collector amplifier circuit which is, of course, based on the much loved common emitter amplifier. The recognizable components are all there. $L_1 C_1$ form the tuned circuit load that will have maximum impedance at resonance. This in turn will give maximum voltage gain (A_v) at the resonant frequency (f_o).

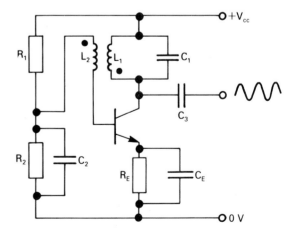

Fig. 7.3   Tuned collector oscillator

R_1, R_2, R_E are the components that set the d.c. or quiescent bias voltages and currents.

Note   The base bias for the transistor is provided by R_1, R_2 and is via the winding L_2, this will have minimal effect since the d.c. resistance of such a coil will be very low. C_E is the usual bypass or decoupling capacitor (that prevents a.c. negative feedback occurring under operating conditions, thus ensuring full a.c. gain). C_3 is a coupling capacitor that blocks d.c. while allowing only the a.c. signals to appear at the output.

Operation

The transistor itself provides 180° phase shift from base to collector so a further 180° of phase shift must be provided. This is achieved by the winding L_2 which is inductively coupled to L_1 (it is a transformer). *Note* the *spot* convention used, indicates that the direction of the two windings is arranged to be in anti-phase, giving 180° phase shift between L_1 and L_2. The total phase shift from input to output and back to input is now 360° so positive feedback has been achieved.

The tuned circuit will ensure that feedback in any quantity will only occur at one frequency, the resonant frequency of the LC circuit. This can be approximated using the formula

$$f_o = \frac{1}{2\pi\sqrt{LC}} \text{Hz}$$

Note   This is the formula for series resonance. A parallel tuned circuit is used that may resonate at a different frequency if the coil resistance is other than negligible (see Chapter 2 – tuned amplifiers). For practical purposes, however, this approximation can be used and provided the overall loop gain $\beta A = 1$, oscillations will build up and be sustained at the frequency f_o.

THOUGHT

What is the purpose of C_2? R_1 and R_2 provide the d.c. base bias for the transistor, oscillations are fed to the base of the transistor from L_2. The oscillations occurring in L_2 will also appear at the junction R_1, R_2 and will upset the d.c. biasing. To prevent this happening a decoupling capacitor (C_2) shorts any oscillations down to the 0 V line thus keeping the biasing constant.

Circuits of this type are usually employed where high frequency sine waves are required. By replacing C_1 with a variable capacitor this oscillator will provide an adjustable output frequency.

The Phase Shift Oscillator

Study Fig. 7.4 and see that once again it has been implemented using the common emitter configuration. R_4, R_5, R_E provide d.c. biasing for the transistor with R_1 as the load resistor. C_E provides decoupling, preventing a.c. negative feedback. C_4 is the coupling capacitor that blocks d.c. allowing only the oscillations to appear at the output.

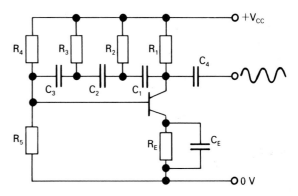

Fig. 7.4   The RC or phase shift oscillator

Operation

It is easy to get confused and say that a CR network has not got a resonant frequency and of course this is true, but in this circuit the capacitors and resistors provide phase shift and cause positive feedback at only one frequency. The transistor itself provides 180° of phase shift from base (input) to collector (output) so a further 180° of phase shift is needed to fulfil the requirements for positive feedback. You will recall from your knowledge of electrical principles that the CR circuit (Fig. 7.5) provides phase shift between V_{in} and V_R or V_C.

If the output voltage is taken to be the voltage across the capacitor (V_C) it will lag the input voltage (V_{in}) by θ° while if V_R is used as the

PRACTICAL INVESTIGATION *14*

The Phase Shift Oscillator

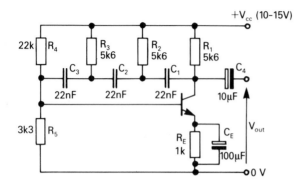

Equipment
D.C. power supply
Oscilloscope (CRO)
General-purpose n-p-n Si
transistor
3 × 5k6 resistor
3 × 22 nF capacitor
1 k, 22 k, 3k3 resistors
100 μF, 10 μF capacitor

Method
1   Build the circuit shown above.
2   Using the CRO monitor V_{out} and check that the circuit is oscillating. (If it is not — *tweak* the power supply voltage up a little.)
3   Measure and record the amplitude and periodic time of the output waveform.
4   Note the effect that varying the power supply voltage has on the output signal.

Results
1   Calculate the frequency of the output signal using your measured periodic time.
2   Calculate the theoretical output frequency using

$$\frac{1}{2\pi RC\sqrt{6}} \text{ Hz.}$$

3   Account for any discrepancies between the theoretical and actual output frequencies.

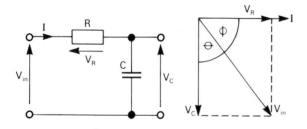

Fig. 7.5   Phase shift provided by a single CR network

output voltage it will lead V_{in} by $\phi°$. Whichever output voltage arrangement is used the maximum theoretical phase shift that can be achieved will be 90° *BUT* at 90° the attenuation will be infinite, i.e. V_{out} will be zero! Therefore more than two networks will be required.

To achieve 180° of phase shift three CR networks are used, each providing 60° phase shift. From Fig. 7.4 (R_1, C_1), (R_2, C_2), (R_3, C_3) are the phase shifting components and since each part must provide 60° it follows that $R_1 = R_2 = R_3$ and $C_1 = C_2 = C_3$.

The CR networks will provide this additional 180° of phase shift at one frequency only and so make the circuit oscillate. This frequency is given by the equation

$$f_o = \frac{1}{2\pi RC\sqrt{6}} \text{Hz}$$

The capacitors and resistors, while providing the required phase shift, do attenuate the signals considerably, such that the feedback achieved (βV_{out}) is only $\frac{1}{29} V_{out}$. The transistor must make up for this attenuation and have a voltage gain $A_v = 29$, only then will the loop gain requirements $A_v\beta = 1$ be satisfied and oscillations maintained.

This is a very useful circuit for providing low frequency sine waves at a single frequency. It is not practical to make the output frequency variable since to do so would require altering all three resistors (or capacitors) in step. Perform Practical Investigation 14 and check the operation yourself.

The Wien Bridge Oscillator

This is another type of RC oscillator that can be designed to provide a high-quality low-distortion sine wave output signal over a very large frequency range.

The Wien bridge network

It is this circuit that controls the output frequency by determining the frequency at which positive feedback occurs. Examine the circuit of Fig. 7.6.

For simplicity let $R_1 = R_2$ and $C_1 = C_2$.

Any purely resistive circuit offers zero phase shift at all frequencies but once capacitors are introduced into a circuit phase shifting will occur that is dependent upon the frequency. The Wien bridge consists of a series CR network in series with a parallel CR network. The input signal (V_{in}) is applied across the complete circuit and the output signal (V_{out}) taken from across the parallel combination.

Due to the phase shift provided by both the networks the voltage V_{out} will be *in-phase* with

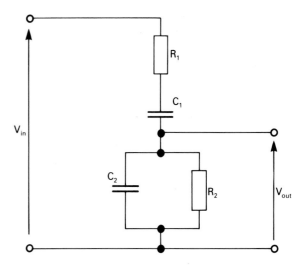

Fig. 7.6   The Wien bridge network

the applied voltage at one frequency only! This frequency (f_o) is given by:

$$f_o = \frac{1}{2\pi RC} \text{Hz}$$

At this frequency there is zero phase shift between V_{in} and V_{out}. However, attenuation will occur such that

$$\frac{V_{in}}{V_{out}} = 3 \; or \text{ if you prefer } \; V_{out} = \frac{1}{3} V_{in}$$

The Wien bridge network can be connected across the output of an amplifier as shown in Fig. 7.7. The voltage developed across the parallel CR network is used as the feedback signal to the amplifier's input. There are two major requirements that must be satisfied if the circuit is to work properly:

1   The amplifier must be non-inverting (since the network provides zero phase shift at f_o).
2   The feedback voltage (βV_{out}) will be $\frac{1}{3} V_{out}$ so the amplifier must have a voltage gain (A_v) of 3 (to ensure that the overall loop gain will be unity ($A_v\beta = 1$)).

This can be achieved in practice by using a two-stage inverting amplifier ($180° + 180° = 0°$) or by applying the positive feedback to the non-inverting input of an operational amplifier.

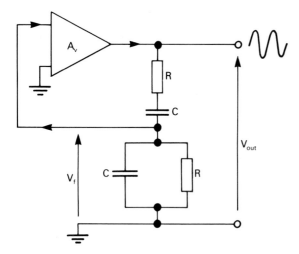

Fig. 7.7   The Wien bridge oscillator

The Wien bridge oscillator using an operational amplifier

Fig. 7.8 shows the complete circuit with the bridge connected across the output of the amplifier and the feedback voltage applied to the non-inverting input. This meets the positive feedback requirements but the amplifier must have a gain A_v of 3 to make up the losses of the feedback network. The gain of the op-amp is determined by the resistors R_f and R_{in} — *Now be*

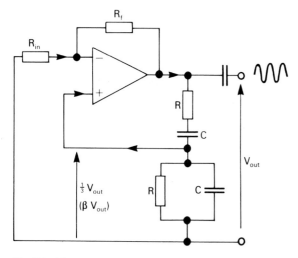

Fig. 7.8   The op-amp Wien bridge oscillator

careful!! Using op-amp theory it is tempting to say that to achieve $A_v = 3$, R_f must be 3 times greater than R_{in}; but this is only true of an inverting amplifier, and we can see that this is operating as a non-inverting amplifier so the correct equation will be:

$$A_v = 1 + \left(\frac{R_f}{R_{in}} \right)$$

The actual gain is critical, if A_v is less than 3 the circuit will not oscillate but, if it is much greater than 3 the output waveform will distort becoming non-sinusoidal. It is usual to make R_f variable and adjust it for the best output.

THOUGHT

Can this oscillator be modified to provide a variable output frequency? — Yes. Simply by substituting a two-gang potentiometer (two variable resistors ganged together) for the resistors in the Wien bridge.

By following the next practical investigation you can examine the operation for yourself.

Frequency Stability of Oscillators

This may at first be thought of as a strange consideration since the basis of an oscillator is an *unstable* amplifier! The stability of an oscillator refers to how well the output frequency and amplitude remain constant. It is clearly undesirable that the frequency should *drift* about or change from its specified output. There are applications where even small variations are disastrous! A radio transmitter emits a carrier signal that must be constant in order for receivers to tune to it. Control systems, clocks, computers, medical equipment etc., often use an oscillator to supply a frequency that determines the timing accuracy. From these examples there appears to be a very definite need for oscillators that have both long-term and short-term stability. The factors that are likely to cause a *drift* in the frequency of oscillations include:

1   Temperature variations.
2   Power supply variations.

PRACTICAL INVESTIGATION **15**

The Wien Bridge Oscillator

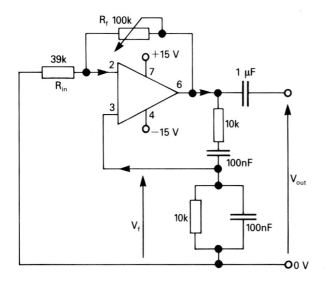

Equipment
Dual power supply
Oscilloscope (CRO)
741 op-amp
2 × 10 k resistors
39 k resistor
100 k variable resistor
2 × 100 nF capacitor
1 μF capacitor
Breadboard

Method
1   Build the circuit as shown.
2   Monitor V_{out} with CRO.
3   Adjust R_f until oscillations occur and an undistorted output is obtained.
4   Measure and record the periodic time and amplitude of the output signal.
5   Measure and record the amplitude of the feedback signal V_f.
6   Observe what happens to the output signal when the gain (A_v) is

a)   increased   b)   reduced.

by varying R_f.

Results
1   Calculate the output frequency from your measurements and compare it to the theoretical value.
2   Calculate the actual gain of your amplifier by using your measured values of V_f and V_{out} (since V_f is the input signal to the amplifier $A_v = V_{out}/V_f$).

3   Changes in component values — particularly in the frequency determining network.
4   Changes in oscillator output current.

These can be minimized by the use of high quality components and careful circuit design but today it is possible to use a piezoelectric crystal to control the frequency of the oscillator.

The piezoelectric crystal

While many people may not have heard of the piezoelectric crystal I am sure most people have a passing familiarity with the term *quartz crystal* since it appears in the name of many domestic articles — the quartz watch and clock — the quartz controlled video recorder and turntable. Even washing machines and vacuum cleaners now feature *quartz* control as one of the selling features. *So what is a piezoelectric crystal?*

The piezoelectric effect

Certain crystals (and quartz is one of these) have the characteristic that when pressed they generate a voltage (piezo is Greek for press!). If such a crystal is struck a *blow* it will generate a voltage waveform as indicated in Fig. 7.9.

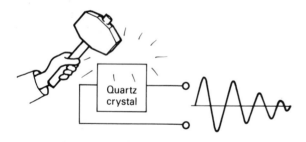

Fig. 7.9   The piezoelectric effect

An important point here is that although the amplitude of the waveform is diminishing, the frequency is constant. This is because the crystal having been struck is vibrating at its own natural frequency — in the same way that a wine glass will emit its own characteristic note when *flicked* with a finger.

The natural frequency of the crystal will be determined by its physical size, i.e. the smaller

and thinner it is the higher will be its natural frequency. Incidentally, it is this voltage characteristic of the piezoelectric crystal that has revolutionized the lighter and igniter industry in the past few years. The modern gas igniter system often uses a small spring-loaded hammer to strike a tiny piezoelectric crystal, and by so doing generate several kilovolts which jumps a spark gap, lighting the gas. The phenomenon is also reversible, i.e. apply a varying signal to a crystal and it will vibrate at its natural frequency. If the applied signal and the crystal have the same frequency the vibrations will be maximum!

The piezoelectric crystal has exactly the same electrical characteristics as a tuned resonant circuit. Its BS symbol and equivalent circuit are shown in Fig. 7.10 (a) and (b), respectively.

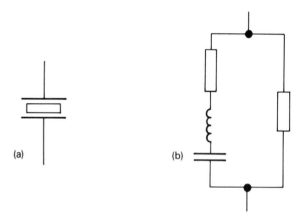

Fig. 7.10   (a) BS symbol for a crystal; (b) Equivalent circuit

The crystal behaves like a very high Q circuit designed to resonate at one specific frequency. The actual crystal used has a temperature coefficient of almost zero, i.e. its characteristics are virtually unchanged by temperature variations. This coupled with the fact that it remains unaffected by ageing, means that frequency stabilities of up to 1 part in 10^7 are easily obtained (± 1 Hz in 10 MHz).

Practical crystal-controlled oscillator circuits

Once it is appreciated that the crystal can be used as the frequency determining network in an oscillator it is simply a matter of replacing the usual network with a crystal, or modifying the circuit so that the crystal acts together with the frequency network to provide stability. Typical circuits are depicted in Figs 7.11 and 7.12.

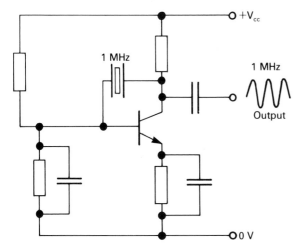

Fig. 7.11   Pierce crystal controlled oscillator

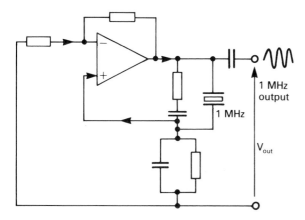

Fig. 7.12   Crystal controlled Wien bridge oscillator

Although it is theoretically possible to design crystals to operate at any frequency, very low frequency crystals would be very large and consequently both 'lossy' and expensive. While very high frequency crystals would be too fragile to use. For these reasons crystals tend to be available from a few kHz up to about 10 MHz. Where higher frequencies are required, a crystal is used but is encouraged to vibrate at a harmonic

of its natural or fundamental frequency, thus allowing larger more robust crystals to provide outputs of up to about 100 MHz.

For use in clocks, watches and other timing circuits it is normal practice to employ a high frequency crystal oscillator, e.g. 4.19430 MHz and digitally divide the frequency repeatedly to obtain hours, minutes, seconds, $1/10$ of seconds etc.

Oscillator Review

1   An oscillator is a circuit that produces an output frequency with no external input signal.
2   An amplifier will produce oscillations provided the following requirements are met:

   (a)   d.c. power
   (b)   positive feedback: $0°$ or $360°$ overall phase shift.
   (c)   a frequency determining network.
   (d)   sufficient gain to maintain oscillations (overall loop gain of unity $\beta A_v = 1$).

3   In order to oscillate, an inverting amplifier will require an additional $180°$ phase shift.
4   A non-inverting amplifier requires no additional phase shift from output back to input.
5   A parallel tuned circuit can be used as a frequency determining network with the output frequency approximated by

$$f_o = \frac{1}{2\pi\sqrt{LC}}\, \text{Hz.}$$

6   A resistor-capacitor network can be used for frequency determination by ensuring positive feedback occurs at one frequency only.
7   An RC or phase shift oscillator uses an inverting amplifier and 3 CR phase shifting networks each providing $60°$ additional phase shift. The frequency is given by:

$$f_o = \frac{1}{2\pi RC\sqrt{6}}\, \text{Hz}$$

8   The losses in a phase shift oscillator require the amplifier gain to be a minimum of 29 to create a loop gain of unity.

9   A Wien bridge oscillator is an RC oscillator in which the RC network produces zero phase shift at the operating frequency.

$$f_o = \frac{1}{2\pi RC} \, \text{Hz}$$

10 The Wien bridge must be used with a non-inverting amplifier or two inverting amplifiers connected in cascade.

11 A minimum gain of 3 is required in the Wien bridge circuit in order to create unity loop gain.

12 Frequency stability of an oscillator refers to its ability to produce a constant frequency output.

13 A piezoelectric crystal can be used to control the frequency of an oscillator.

14 The piezoelectric effect refers to a crystal's ability to produce an output voltage when vibrated, conversely if a voltage is applied to the crystal it will itself resonate.

15 A crystal acts like a very high Q tuned circuit. It has a very low temperature coefficient and will easily provide a stability of 1 part in 10^7 Hz.

16 There are a number of other forms of oscillator which are adequately dealt with in other textbooks.

8

Thyristors and Triacs

Before we consider this device let us recap on the humble diode. The diode is a two-layer, single-junction semiconductor device that acts as a *one way valve* to current flow. It has two states: forward and reverse biased, as shown by the characteristic in Fig. 8.1.

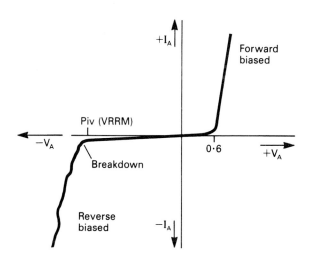

Fig. 8.1 Diode characteristic

When reverse biased, a diode *blocks* the flow of current (apart from a tiny amount of leakage) until the reverse bias voltage reaches such a high level that breakdown occurs; after this the device is useless. When a silicon diode is forward biased, appreciable current will only start to flow when the anode is about 0.6 V more positive than the cathode. The diode will now conduct.

So when forward biased a diode has a low resistance and current will flow through it — *conducting state*. When reverse biased a diode has a very high resistance and very little current flows — *blocking state*.

The thyristor has the alternative name of *Silicon Controlled Rectifier* (SCR). This, in fact, describes its function very well since it acts like a diode that can be triggered into conduction. To enable this it has a trigger electrode called the *Gate*. The BS symbol is shown in Fig. 8.2 and bears a similarity with the diode symbol.

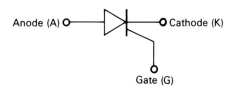

Fig. 8.2 BS symbol for a thyristor

This triggering facility means that the thyristor has three possible states when compared to a diode's two states, they are:

- The reverse blocking state, Fig. 8.3(a), the device is reverse biased so no current flows.
- The forward blocking state, Fig. 8.3(b), although forward biased no current flows because a trigger signal has not been applied to the gate.
- The conducting state, Fig. 8.3(c), the device is forward biased and a trigger signal has been applied so current (I_a) now flows. This is the anode current.

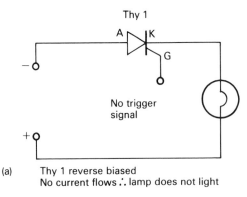

(a)   Thy 1 reverse biased
No current flows ∴ lamp does not light

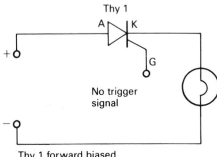

(b)   Thy 1 forward biased
but no trigger signal ∴ no current flows

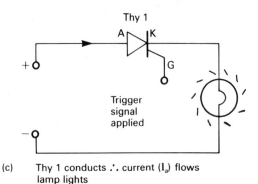

(c)   Thy 1 conducts ∴ current (I_a) flows
lamp lights

Fig. 8.3   (a) Reverse blocking; (b) Forward blocking; (c) Forward conduction

Thyristor Triggering and Operation

To develop an understanding of the way this device performs it is best to study the characteristics shown in Fig. 8.4.

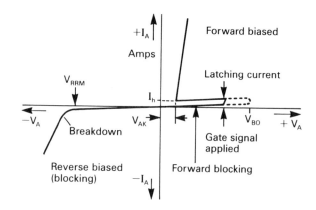

Fig. 8.4   Thyristor characteristics

You will see that when reverse biased, the characteristic is exactly like a diode. When forward biased, however, no current flows until the gate trigger signal is applied; *turn on* now occurs and forward conduction takes place.

The gate trigger signal

A positive pulse of low amplitude, e.g. 2 V, 10 mA (depending on the device), is all that need be applied to the gate to initiate turn on. Once this has happened, however, current will continue to flow regardless of the state of the gate electrode.

THOUGHT _____

Does this mean that once triggered into conduction the thyristor cannot be turned off by the gate? — This is precisely what it means, whatever is done to the gate — make it negative, positive, pulse it, earth it, cut it off completely, etc., will have no effect on the thyristor at all which will continue to conduct happily!

How can it be turned off? If you look at the characteristic of Fig. 8.4 you will see a forward current I_h indicated. This is the *holding current*. It is the minimum current flow required to keep the thyristor turned on. If the current falls below this value the device will turn off. A conducting thyristor can be turned off in two ways:

1   The current through it can be reduced to below I_h (Fig. 8.5).

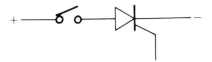

Fig. 8.5 Turn off by interrupting the anode current

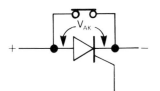

Fig. 8.6 Turn off by reducing V_{AK} to zero

electromechanical arrangement can be easily replaced by the thyristor circuit of Fig. 8.8.

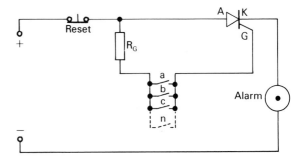

Fig. 8.8 Burglar alarm using a thyristor

2   The voltage across the thyristor can be reduced to zero (Fig. 8.6). Thus causing I_h to fall to zero.

Once turn off has been achieved the device will again be in the forward blocking state until a further gate trigger pulse is applied when it will again turn on. In this respect the thyristor can be considered as a *self-latching relay* since its operation under d.c. conditions is the same. Consider the burglar alarm circuit of Fig. 8.7.

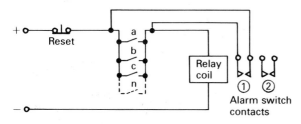

Fig. 8.7 Burglar alarm using a relay

Switches a–n are normally open intruder entry transducers, e.g. reed switches on door and window frames or pressure pads beneath carpets etc. If any one of these switches is momentarily closed the relay becomes energized and the two sets of contacts close. Contact (1) overrides the intruder switches and the relay *latches on*, the alarm will continue to sound until the reset button interrupts the supply. If the intruder switches are in the normally open position the circuit will now be back in its *ready* state. This

If any of the switches a–n is momentarily closed the thyristor will receive a gate signal and since it is forward biased by the d.c. supply it will conduct, thus sounding the alarm until the *reset* is operated, manually.

THOUGHT

All this is fine but what is the advantage?
A thyristor is a contactless switch with no moving parts, there is nothing to corrode or jam, and no cause for concern about contact wear or burning. In addition to these desirable facts is the realization that the thyristor like the diode is available with high current and voltage ratings. The gate signal required to turn it on is still very small, i.e. a 100 A device can be turned on with a 3 V gate signal of no more than 100 mA. This makes it very useful as a control device.

The two-transistor analogy

The thyristor is a four-layer (p-n-p-n) silicon device (Fig. 8.9(a)) and to help appreciate how it operates it can be considered as two complementary transistors connected as shown in Figs 8.9(b) and (c).

T_1 is a p-n-p transistor and T_2 an n-p-n transistor. When A is positive with respect to K the device is forward biased, conduction will not occur because, although the emitter base junction of T_1 is forward biased, the collector base junction of T_2 is reverse biased. If a positive potential is applied to G base current (I_{B2}) will flow in T_2 which will cause collector current to flow through T_2. Current can now flow from A to K and once this flow

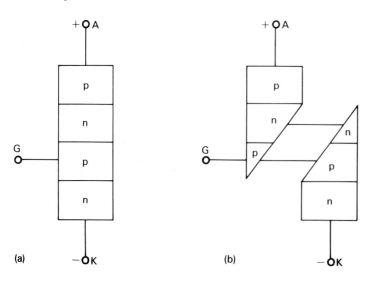
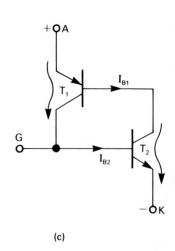

Fig. 8.9 (a) Thyristor construction; (b) Thyristor construction;
(c) Two transistor analogy

is established base current I_{B1} will be flowing in T_1, this will result in collector current flowing in T_1 and this collector current will form the base current (I_{B2}) for T_2. Each transistor now supplies the base current for the other and so saturation very quickly occurs. A perfect closed loop now exists and the gate signal is surplus to requirements.

Note   This is *NOT* a practical circuit since in this analogy the anode current is the transistor base current.

The operation of the device can be checked by performing Practical Investigation 16.

Other methods of turning a thyristor 'ON'

So far only gate triggering of a forward biased thyristor has been considered. There are two other methods by which a thyristor can be *turned on*:

1   The forward bias voltage ($+V_A$) can be increased to a point where *forward breakover* takes place and conduction occurs. This is shown as V_{BO} on the characteristic of Fig. 8.4.
2   If a positive going voltage waveform is applied between anode and cathode then normally the device is forward biased but non-conducting. If this waveform has a very

fast rise-time, e.g. 70 V/μs, then the internal capacitances that exist between the layers can allow this signal to be fed through to the gate thus triggering the device into conduction.

Both of these methods result in unreliable triggering and are, in fact, actively avoided in design as will be seen from the following.

Choosing a Thyristor

Like a diode the application will determine the specifications or ratings that will be required of the chosen device. To help with the selection, manufacturers quote certain data relating to thyristors:

V_{RRM} — The *repetitive peak reverse voltage*, i.e. the maximum reverse bias voltage that can be applied without breakdown occurring.

$I_T(AV)$ — The maximum mean anode current that can be passed through the device without damage occurring.

I_H — Holding current, the minimum anode current required to keep the device turned on.

PRACTICAL INVESTIGATION 16

Basic Thyristor Operation

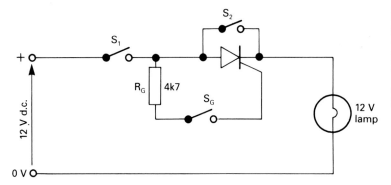

Equipment
d.c. power supply
12 V lamp (up to 5 W)
Medium power thyristor, e.g.
TIC106B
4k7 resistor
3 switches

Method
1 Connect the circuit as shown above.
2 Set switches S_2, S_G into the *open* condition.
3 Switch on the power supply and adjust for 12 V.
4 Close S_1.
5 Momentarily close S_G and observe that the lamp lights and remains on.
6 Reset the circuit by opening S_1.
7 Repeat the latching operation and reset using S_2.
8 Disconnect the gate of the thyristor so that it is *floating*.
9 With the d.c. supply on and S_1 closed investigate what happens when the gate terminal is handled.

Results
1 What is the purpose of the gate resistor R_G?
2 How sensitive is the gate?

V_{GT} — The minimum instantaneous gate trigger voltage required to turn the device on.

I_{GT} — The minimum instantaneous gate trigger current required to turn the device on.

V_T — The voltage drop across the device when it is conducting.

If you study the data sheet on page 156 you will see that the data provided is the minimum required for selection of a device. If more detailed information is needed, this must be obtained from the manufacturer of the specific thyristor.

Practical Investigation 17 allows you to measure some of the important parameters for yourself.

PRACTICAL INVESTIGATION 17

Thyristor Characteristics

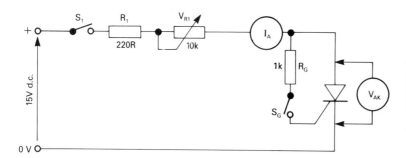

Equipment
d.c. power supply
Voltmeter (20 V)
Ammeter (200 mA)
220R 1 k resistor
2 switches
Medium power thyristor, e.g.
TIC106B
10 k potentiometer

Method
1   Connect the circuit as shown.
2   With S_1 and S_G open connect the d.c. power supply and adjust to give 15 V. Close S_1.
3   Set V_{R1} to its minimum value and record the readings of the anode current (I_A) and the voltage across the thyristor (V_{AK}).
4   Momentarily close S_G — the thyristor is now triggered.
5   Record the values of I_A and V_{AK}.
6   To determine the holding current (I_h) increase the value of V_{R1} slowly while observing I_A as it reduces. At some point the thyristor will turn off. I_h is the current indicated just prior to turn off occurring.

D.C. Operation of the Thyristor

Once the basic operation of the thyristor has been understood there appear to be certain limitations, chiefly centred around the problem of turning the thing off! The advantage is clearly that a high current can be switched on with a tiny gate signal but to stop conduction, the supply to the anode must be broken, or alternatively a switch connected between anode and cathode can be closed momentarily so that the current falls below I_h (Fig. 8.10). With both these methods the switches must be capable of passing the full operating current. This is pointless since the resulting arrangement will consist of a minia-ture push button for switching *ON* and an enormous high current switch for turning the current OFF!

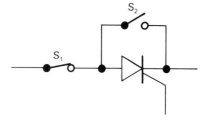

Fig. 8.10   Methods of turning a thyristor off

A solution to this problem is to use a com-*mutating capacitor* as shown in Fig. 8.11.

PB1 and PB2 are push buttons with normally open contacts. When PB1 is momentarily closed, the thyristor receives a gate pulse, the device latches *on* and current flows in the load. The anode (A) is connected to one plate of capacitor C and since Thy1 is conducting, will be at a low

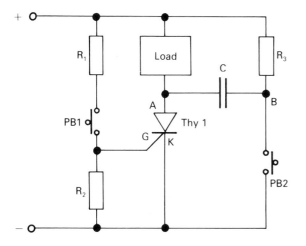

Fig. 8.11   Commutating capacitor

potential. The other plate of C is connected to point B and so the capacitor will charge via R_3 to the supply rail voltage. If PB2 is closed the stored charge in C will be connected directly across the thyristor, this will reverse bias the thyristor and turn it off.

Although this is an obvious improvement the next step is to arrange for the *off* switch PB2 to be electronic. Fig. 8.12 shows a thyristor used to turn *off* a thyristor.

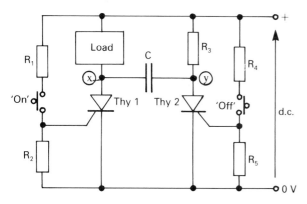

Fig. 8.12   Electronic switching

At *switch on*, both thyristors are *off*.

1   If the *on* button is pressed Thy1 triggers and current flows in the load.
2   Point (x) is now low but point (y) is high (Thy2 is off so no current flows through R_3) C charges via R_3 to the supply voltage.

3   If the *off* button is pressed Thy2 is triggered and point (y) now becomes *low*.
4   C is now connected across Thy1, its stored charge reverse biases Thy1 turning it off.
5   Point (x) is now *high* so C will charge to the supply potential via the load.
6   If the *on* button is again pressed Thy1 triggers, (x) becomes *low* and C discharges across Thy2 turning it off.

It may be convenient to replace the push buttons by electronic pulses that will provide the gate triggering signals.

Problems

This circuit works very well in practice but there are a few weak links:

1   If after the supply is connected the *ON* and *OFF* buttons were pressed simultaneously both thyristors would trigger and remain on. Points x and y would both be low so the capacitor may not charge. It may not now be possible to turn off the circuit.
2   The stored charge in C is critical. It must be large enough to turn off the conducting thyristor. For this reason C must be a high value capacitor and since it charges in alternate directions, i.e. x is negative with respect to y and then y is negative with respect to x it cannot be a polarized capacitor (electrolytics are no good here!).
3   The charge in C will take time to accumulate, this will limit the speed at which the thyristor can be switched. If only a short time is allowed between *ON* and *OFF* operations there may be insufficient charge in C to turn off the thyristor.

The next practical investigation allows you to build and test a similar circuit.

A.C. Control Using the Thyristor

When used with alternating current the thyristor really comes into its own as a very useful control device. It has always been desirable to be able to control the current flowing in a load in order to achieve such things as, light dimming, motor speed control and the temperature control of a heater or oven. Temperature control is possible by switching the load current on and off via a thermostat. If it was required to vary or reduce the load current for such applications as lamp dimming or motor speed control then this was achieved using large variable resistors called rheostats. This method had a major disadvantage in that the load current would have to flow through the variable resistor; the resistor would have to be capable of passing this current and dissipating the resulting heat that would be generated. The result was a monstrously large variable resistor producing wasted power and creating a low efficiency system. Consider Fig. 8.13.

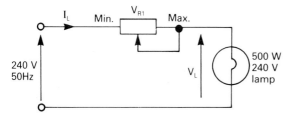

Fig. 8.13   Lamp dimmer using a rheostat

When connected to 240 V a.c. the lamp current (I_L) is given by $500/240 = 2.08$ A. If V_{R1} is set to minimum the lamp will receive the full supply voltage and glow with full brilliance. With V_{R1} set to any other value the circuit of Fig. 8.13 is essentially two resistors in series with the result that the current (I_L) will be reduced and hence the lamp will glow less brightly. However, power will now be dissipated in the resistor V_{R1} as well as in the lamp.

Example

If the lamp resistance (R_L) is 115 Ω and V_{R1} has a resistance at its mid-point setting of 115 Ω calculate:

1   The current flowing in the load.
2   The load voltage.
3   The power in the load.
4   The power in the resistor V_{R1}.

1   Load current

$$I_L = \frac{V_s}{R} = \frac{240}{115 + 115} = 1.04 \text{ A}$$

2   Load voltage
$$V_L = R_L . I_L$$
$$R_L = 115 \text{ }\Omega$$
$$I_L = 1.04 \text{ A}$$
$$V_L = 115 \times 1.04 = 119.6 \text{ V}$$

3   Load power $= I_L V_L = 1.04 \times 119.6 = 124.3$ W
4   Power in $V_{R1} = I^2 R = (1.04)^2 \times 115 = 124.3$ W

Result — both lamp and resistor dissipate power. You can see from this example that control of an a.c. load using this method is not at all efficient.

The thyristor, as a control device, offers the following advantages:

• It can be triggered at any point in the positive half cycle of an a.c. wave.
• Once triggered it will remain on until the supply passes through zero, and because the supply is a.c. it does this every half cycle.

To see how this can help with a.c. control, study the circuit of Fig. 8.14 together with the waveforms of Fig. 8.15.

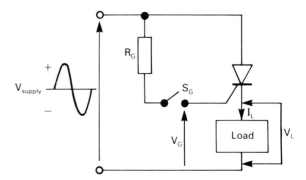

Fig. 8.14   Basic a.c. control circuit

Supposing the gate switch S_G was kept closed, the thyristor would have a positive gate voltage whenever the supply was positive, Fig. 8.15(a). This means that it will act exactly like a diode and provide half wave rectification so load current and voltage waveforms will be as shown in Fig. 8.15(b).

Now supposing S_G was closed momentarily at say 45° after the start of each positive half cycle! The gate waveform would then be a positive pulse as shown in Fig. 8.15(c). This would trigger the thyristor giving the load current and voltage waveforms as shown in Fig. 8.15(d).

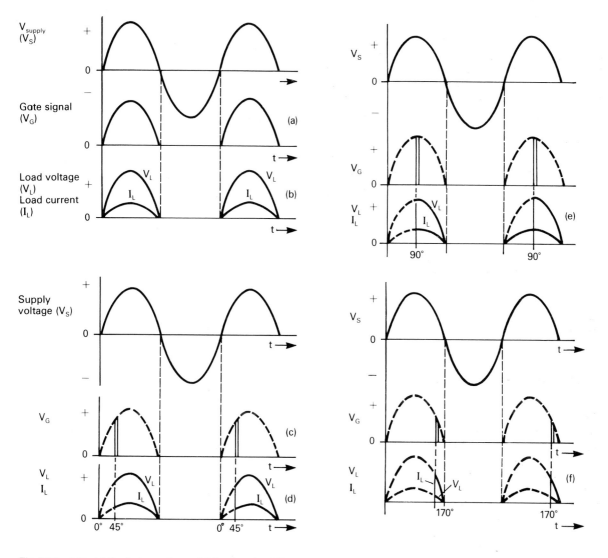

Fig. 8.15 (a) to (d) Control waveforms; (e) S_G closed at 90°; (f) S_G closed at 170°

Now varying the point at which switch S_G is closed during the positive half cycle of the supply voltage will determine exactly when the thyristor will conduct, and thus for how long current will flow in the load. Figs 8.15(e) and (f) show S_G closed at 90° and 170°, respectively.

This is extremely exciting because it shows that we can control the current flowing in the load by deciding when to turn the thyristor on.

THOUGHT

What about turning it off? — No need to worry about this — as soon as the supply passes through zero the thyristor turns off and remains off until it is triggered again during the next positive half cycle.

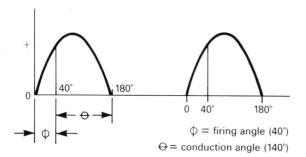

Fig. 8.16 Firing and conduction angles

Obviously the switch S_G cannot be operated manually, so an electronic trigger circuit is used to fire the thyristor at the required point in the positive half cycle. This firing or trigger point is referred to as the *firing* or *delay* angle, φ. After firing, the thyristor conducts for the remainder of the half cycle and this is the *conduction* angle, θ (Fig. 8.16). These parameters are always quoted in degrees. There are 360° in one complete cycle regardless of its frequency.

Phase control of the thyristor

We are now in the position where to control the current flowing in an a.c. load a trigger signal must be applied to the gate of the thyristor at the same point in every half cycle in order to maintain the same conduction. This appears to be a complicated problem but in reality can be solved by using a capacitor/resistor network as shown in Fig. 8.17.

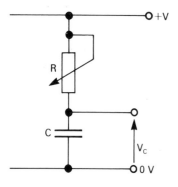

Fig. 8.17 CR network

A thyristor requires a certain level of gate voltage to trigger. Capacitor (C) takes a certain time to charge via resistor (R). Therefore use the voltage (V_C) across the capacitor to act as the gate trigger voltage. If the charging resistor R is variable, the time taken for the capacitor to charge to the trigger level can be altered. Let us now put these various ideas together to produce the complete a.c. phase control circuit of Fig. 8.18.

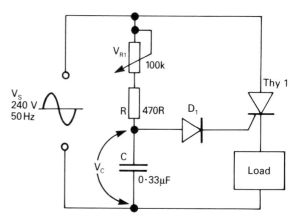

Fig. 8.18 Phase control of a thyristor

The principle of operation of this circuit is that C charges via V_{R1} and R during the positive half cycles of the main input voltage.

When the charge on C is high enough to overcome the combination of the diode (D_1) forward voltdrop and the thyristor gate threshold voltage, Thy1 will trigger, and current will flow in the load until the a.c. passes back through zero when Thy1 will turn off. As Thy1 conducts it discharges C. This will be repeated with each positive half cycle. During the negative half cycles the thyristor will be reverse biased and no current will flow in the load.

THOUGHT

What is D_1 for? — This is to provide a threshold or switching level of approximately 0.6 V.

By altering the value of V_{R1} the firing angle can be varied, so current flowing in the load and hence the power can be controlled.

It is worth considering the action of the CR network in a different way.

PRACTICAL INVESTIGATION *18*

D.C. Line Operation of Thyristors

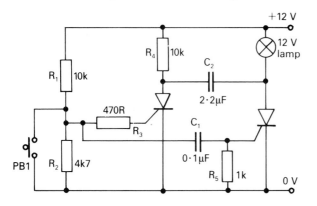

Equipment
Power supply
12 V lamp
2 thyristors, e.g. T1C106
2 × 10 k resistors
470R, 1 k, 4k7 resistors
0.01 μF, 0.1 μF, 0.47 μF,
1 μF, 2.2 μF capacitors
Push button switch with
normally open contacts

Method
1   Build the circuit shown above.
2   Operate the push button (PB1) and observe the operation of the circuit.
3   Replace C_2 with a 1 μF capacitor and note the effect this has on the operation of the circuit.
   With C_2 = 2 μF replace C_1 with a) a 0.01 μF capacitor b) a 0.47 μF capacitor. In each case
   note the effect this has on the circuit operation.

Results
1   Explain why you think the value of C_2 is critical for the correct operation of the circuit.
2   Suggest a method by which this circuit could be converted or extended so that the lamp
   flashes at a specified rate.

The supply signal is a sine wave so the CR network provides a *phase shifting* action — hence the name phase control. The actual firing angle (ϕ) will be determined by the phase angle between the supply voltage (V_s) and the voltage across the capacitor (V_C); this can be estimated using:

$$\text{Tan } \phi \simeq 2\pi fCR$$

but this is only an approximation since attenuation and the thyristor trigger threshold will serve to increase the firing angle. Remember the greater the firing angle the lower will be the load power. Fig. 8.19 shows this relationship.

Perform Practical Investigation 19 and see if you can determine any obvious disadvantages with this circuit.

From your deliberations you may have discovered that the major set-back with this circuit is the fact that the maximum output

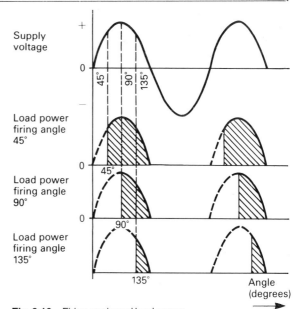

Fig. 8.19   Firing angle and load power

PRACTICAL INVESTIGATION *19*

Phase Control of a Thyristor

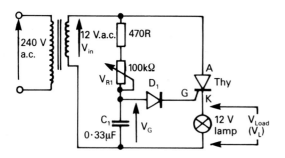

Equipment
Dual beam CRO
240 V: 12 V transformer
12 V lamp (0.5 watt)
Thyristor, e.g. T1C106
100 k variable resistor
470R resistor
Rectifier diode
0.33 μF capacitor

Method
1 Construct the circuit shown above.
2 Monitor V_S and V_L and observe the effect of varying V_{R1}.
3 On a common time scale sketch the waveforms V_S and V_L for minimum, mid-point and maximum settings of V_{R1}.
4 By varying V_{R1} estimate the range over which the firing angle can be varied.

Results
1 What is the main disadvantage with this circuit for the control of a load.

---

consists of half wave rectification. This is certainly limiting and we will deal with the solution in a moment.

The other disadvantage is less obvious, *consider what is happening*: 50 times a second the thyristor is switching *on* and *off*, so the load current is also flowing intermittently. If the firing angle is 90° the thyristor is turning on when the supply voltage is maximum. From your knowledge of electrical *noise* you may have realized that this switching will cause massive interference because of radiated energy. Some form of suppression must be used to minimize this and is usually achieved with an inductor placed in series with the supply to the circuit as indicated in Fig. 8.20.

L_1 consists of a ferrite core with about 20 turns of suitable wire around it. This provides sufficient inductance to attenuate any spikes causing radiation.

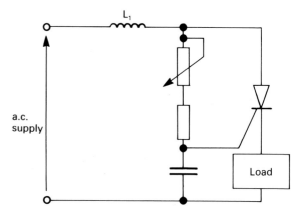

Fig. 8.20   Interference suppression

Solutions to the half wave problem

As the thyristor can only conduct on the positive half cycles one solution is to use full wave rectification to provide unidirectional pulses at twice the input frequency as shown in Fig. 8.21.

The thyristor will now be able to trigger on every half cycle since they are all positive and so full-wave control will now exist.

An alternative would be to use two thyristors connected in reverse parallel so that one would always be able to conduct when the other was reverse biased. This, however, is overcomplicating matters and the modern solution is to use a *Triac* which is effectively two thyristors connected back to back in a single package.

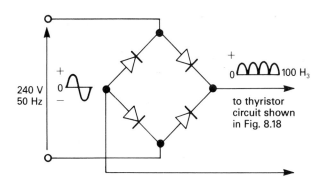

Fig. 8.21　Undirectional supply

The Triac

This device is a *bi-directional thyristor*. This means that it can be triggered into conduction in both directions, Fig. 8.22 shows the BS symbol and Fig. 8.23 the characteristic.

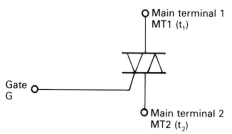

Fig. 8.22　BS symbol for a triac

You will notice that the electrodes of a triac are labelled main terminals MT1 and MT2. This is because since the device operates in both directions the term cathode and anode become meaningless.

Triac triggering

Firing of the triac takes place via the gate. Since it is a bi-directional device it can be triggered into conduction by a negative or positive gate signal. Triac potentials are all considered with respect to main terminal 1 (MT1). This gives the following possible operating situations or modes:

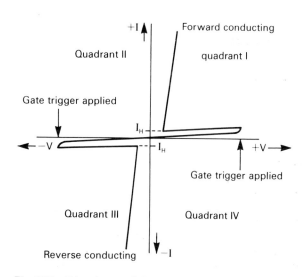

Fig. 8.23　Triac characteristics

MT2 positive with respect to MT1	Gate signal positive　(first quadrant 1+)
MT2 positive with respect to MT1	Gate signal negative　(first quadrant 1−)
MT2 negative with respect to MT1	Gate signal positive　(third quadrant 111+)
MT2 negative with respect to MT1	Gate signal negative　(third quadrant 111−)

Unfortunately the triac is not equally sensitive in all modes!

It is least sensitive in the 111+ configuration (MT2 negative with respect to MT1 triggered by a positive gate signal) so this arrangement is seldom used in practice.

Full wave control using a triac

Fig. 8.24 shows a triac used to control the current flowing in an a.c. load.

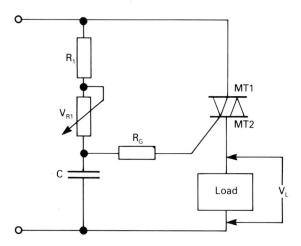

Fig. 8.24   Simple triac control circuit

By studying the waveforms of Fig. 8.25 you can see that control is achieved by firing the triac at the same point in both the positive and negative half cycle. Once triggered the device remains on until the supply passes through zero when it switches off.

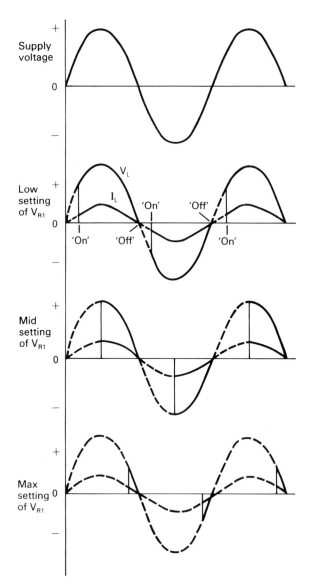

Fig. 8.25   Triac control waveforms

Choosing a Triac

Like all components the triac has maximum specified values of current and voltage that must not be exceeded. The data sheet on page 158 indicates these.

I_T (rms)   The maximum rms current that can continuously pass without damage occurring.

I_{TSM} The maximum (peak) *surge* current that can be tolerated for a very brief period without damage occurring.

I_{GT} The value of gate current required to achieve switch on. *Note* the four possible modes of operation are quoted and that operation in mode $111 + (t_2 - g +)$ is the least sensitive — usually twice the requirement of the other modes.

V_{GT} The value of gate voltage required to achieve switch on.

V_{DRM} (max) The maximum permitted peak voltage. *Note* with the triac the terms forward and reverse do not exist since it is bi-directional.

The Gate Turn Off Thyristor (GTO)

This is similar to a conventional thyristor in that it has three electrodes (Fig. 8.26).

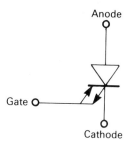

Fig. 8.26   BS symbol for GTO thyristor

Like the thyristor this device must be forward biased and then a positive gate signal will turn it on. Turn off however can be achieved by applying a negative pulse to the gate. The gate turn off thyristor combines the control normally found with a transistor together with the current and voltage capabilities of the thyristor. Let us consider a typical device the BTW58-1300R.

$$I_A \text{ peak} = 6.5 \text{ A}$$
$$V_{RRM} = 1300 \text{ V}$$
$$Turn \; on = \text{(positive gate pulse)}$$
$$V_{GT} = 1.5 \text{ V}$$
$$I_{GT} = 100 \text{ mA}$$
$$\text{minimum gate pulse duration} = 10 \text{ μs}$$
$$Turn \; off = \text{(negative gate pulse)}$$
$$V_{GT} = 15 \text{ V}$$
$$I_{GT} = 100 \text{ mA}$$
$$\text{minimum gate pulse duration} = 1 \text{ μs}$$

From this data you can see that the pulse duration and amplitude requirement differ for

the *ON* and *OFF* operation and because a polarity charge is required the drive and control circuits for use with this device are different to the circuits used with normal thyristors and triacs.

Switching Inductive Loads

So far we have only considered thyristors and triacs controlling the power in purely resistive loads. In reality of course, many loads possess inductance, e.g. motors. We must now consider the special problems that an inductive load presents when switched by a thyristor. Fig. 8.27 shows a thyristor switching a purely resistive load.

The waveforms of Fig. 8.28 show that when the thyristor is triggered the load voltage and current

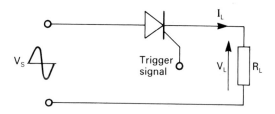

Fig. 8.27   Thyristor switching a resistive load

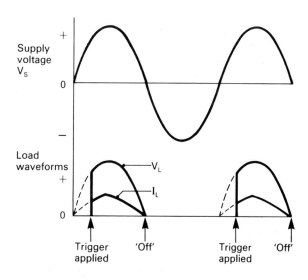

Fig. 8.28   Resistive load waveforms

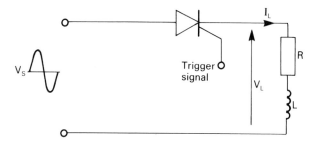

Fig. 8.29   Thyristor switching an inductive load

signals are in phase and consequently the thyristor current will be zero when the load voltage falls to zero. At this point the device will turn off. Fig. 8.29 shows the same thyristor arrangement but this time it is controlling an inductive load.

The load can be considered as having a resistive (R) and inductive (L) component. The inductor will cause the load current to lag the load voltage as indicated in Fig. 8.30.

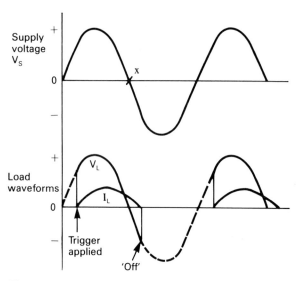

Fig. 8.30   Inductive load waveforms

The effects of this inductive phase lag are two-fold:

1   When the supply voltage (V_s) has fallen to zero (point x) the load current (I_L) is not zero but is still falling. This means that the current flowing in the thyristor will be maintained

above the holding current (I_h) for some while, with the result that the thyristor will continue to conduct for part of the negative half cycle of the supply voltage. When the current flowing in the load finally falls below the holding current the device will turn *off.*

THOUGHT

So what! All this means is that some 'blips' of negative load voltage will occur. Surely this can do no harm since it merely serves to make the circuit more like a full wave rectifier?

With an inductive load this phase lag phenomenon will always occur and is determined by the ratio L/R. The larger this ratio the longer the thyristor will conduct during the negative half cycle. This will mean that there is less time available during each cycle for the thyristor to recover from the effects of conduction. Result — loss of control.

Solution

The problem is that the load current is not zero when the supply voltage is zero. This is due to a back EMF generated by the inductive load maintaining the flow of current (Lenz's Law). This can be corrected by placing a flywheel or commutating diode across the load as shown in Fig. 8.31.

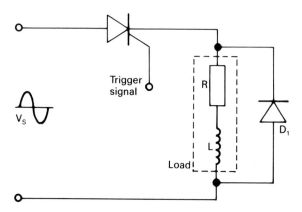

Fig. 8.31   The flywheel diode

When the supply voltage V_s passes through zero and starts to go negative the diode D_1 conducts, allowing the energy stored in the inductor to dissipate in the load, and thus effectively reducing the current through the thyristor to zero allowing it to turn off.

2 Thyristors and triacs tend to be triggered with short duration gate pulses. The load current lags behind the supply voltage and there is a risk that the thyristor current may not reach the latching current by the time the gate pulse is finished.
Result: The thyristor will not turn on! Leading to loss of control.

THOUGHT

Is the latching current the same as the holding current? No! In practice the thyristor requires a minimum current to pass in order to fully turn on. This is the latching current. Once fully conducting, however, the thyristor will remain on until the current falls below a minimum value called the holding current.
The holding current is lower than the latching current

An example will help illustrate this problem.

Example

A thyristor with a latching current of 50 mA is used in the circuit of Fig. 8.32. A gate pulse of 60 μs is used but the thyristor will not turn on.

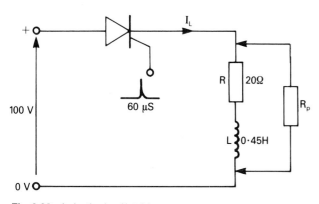

Fig. 8.32 Inductive load latching

The time duration of the gate pulse = 60 μs.
The time constant of the load

$$= \frac{L}{R} = \frac{0.45}{20} = 22.5 \text{ ms}$$

Latching current = 50 mA
Supply voltage = 100 V

From these calculations it can be seen that the time constant of the inductive load is long when compared to the gate pulse duration (22.5 ms vs 60 μs). It is unlikely that the current (I_L) will rise to 50 mA in the 60 μs that the gate pulse is present, so the thyristor will not turn on. This can be shown thus:

The rate that the current will rise is assumed to be linear and will be governed by V/L.

$$\therefore \text{ rate of rise of current} = \frac{100}{0.45} = 222 \text{ A/s}$$

so in 60 μs the current will be $222 \times 60 \times 10^{-6} = 13.3$ mA this is far below the required 50 mA.

Solution

Connect a resistor (R_p) in parallel with the load to ensure that the instant a gate pulse is applied 50 mA flows. See Fig. 8.32.
The value of R_p required can be found in the following way:

$$R_p = \frac{V}{(I_{latch} - I_{actual}) \times 10^{-3}}$$

$$= \frac{100}{(50 - 13.3) \times 10^{-3}}$$

$$= \frac{100}{36.7 \times 10^{-3}} = 2.72 \text{ k}$$

A 2.7 k resistor will ensure that at switch on 50 mA will flow and the thyristor will latch.

Thyristor and Triac Review

1 The thyristor or *Silicon Controlled Rectifier* (SCR) is a device that can be triggered into conduction when forward biased by the application of a positive gate signal.
2 Once triggered the thyristor will continue to conduct regardless of the state of the gate.
3 To *turn off* a thyristor the current flowing through it must be reduced to a level below the holding current (I_h), or the anode voltage reduced to zero.

4   When operated on d.c. the thyristor acts like a self-latching relay.

5   When operated on d.c. *turn off* can be achieved by using a commutating capacitor.

6   To control the a.c. power in a load a thyristor can be triggered by a positive gate signal at any time during the positive half cycle of the a.c. supply voltage.

7   The triggering point is called the firing or delay angle.

8   After firing the thyristor conducts for the remainder of the positive half cycle, turning off when the supply voltage passes through zero.

9   Phase control can be achieved using a CR network to delay the gate trigger signal.

10   The greater the firing angle the lower will be the load power.

11   A single thyristor will give only half wave control when used with an a.c. supply.

12   A triac is a bi-directional thyristor, i.e. it can be triggered into conduction during positive and negative half cycles by a negative or positive gate signal.

13   Thyristor and triac a.c. control circuits generate interference due to the load switching that is occurring. This requires suppression, usually achieved with an inductor placed in series with the load.

14   The gate turn off thyristor (GTO) is a special thyristor that can be turned on and off with a gate control pulse.

15   When switching inductive loads, the load current will lag the supply voltage. This may result in the thyristor:
a) conducting during part of the negative half cycle,
b) not latching when triggered.

16   Inductive load problems can be overcome using a parallel load resistor (R_p) to ensure latching and a flywheel diode to take the anode current to zero at the same time as the supply voltage reaches zero.

9

Thyristor and Triac Triggering

In the previous chapter we have concerned ourselves with the operation of power control devices and simple circuits that allow the firing angle to be varied. So far the thyristors and triacs have been switched by a gate voltage that rises until it reaches the trigger *on* level, at which point the device conducts.

For good a.c. control we rely on the fact that the thyristor is fired at the same point in each cycle. However, the device is presented with a rising gate voltage and will fire when this is at or near the thyristor's minimum gate trigger voltage (V_{GT}). Unreliable triggering may result from this arrangement, with the firing angle varying from cycle to cycle, and device to device. For accurate firing and reliable latching, thyristors and triacs perform best when triggered by a gate pulse rather than an increasing level. One technique used is to place a Zener diode in series with the gate of a thyristor as shown in Fig. 9.1.

The voltage across the capacitor will rise until the Zener breakdown voltage is reached then the thyristor will receive a gate pulse as the Zener conducts. When the Zener conducts the capacitor is discharged. The value of the Zener diode will determine the size of the gate pulse.

For triac triggering a device is required that will provide a negative and positive pulse in order that the triac can be triggered during both the positive and negative half cycle. The component that does just this is the *diac*. Fig. 9.2 shows the symbol and Fig. 9.3 the characteristic.

Fig. 9.2 BS symbol for a diac

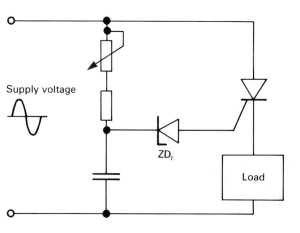

Fig. 9.1 Gate triggering using a Zener diode

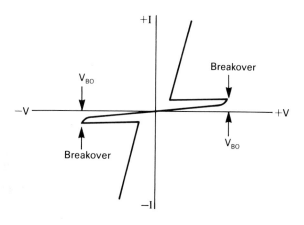

Fig. 9.3 Diac characteristic

The diac is in reality a triggerless triac that is designed to break down at a precise low voltage in both directions, typically 35 V. A typical triac control circuit using a diac is shown in Fig. 9.4.

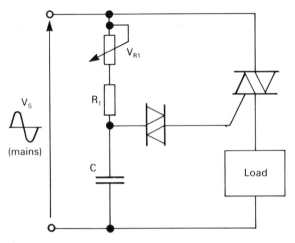

Fig. 9.4 Triac triggering using a diac

During both the positive and negative half cycle the triac will only receive its gate pulse when the voltage across the capacitor reaches the diac breakover or threshold voltage V_{BO} of 35 V. This simple arrangement provides very reliable triggering and is widely used for *lamp dimmer* circuits that can be fitted in place of standard mains light switches.

The Unijunction Transistor (UJT)

This is a semiconductor device that is frequently used for triggering power control devices. It also finds application in oscillator and timing circuits.

UJT construction and operation

Fig. 9.5 shows the physical construction of the UJT and the circuit symbol.

The device consists of a bar of 'n' type silicon with base electrodes at both ends (B_1, B_2). These electrodes are ohmic connections, i.e. they are not junctions just simple electrical terminations. A p-type region is formed in the bar by inserting an aluminium emitter electrode at this point. A

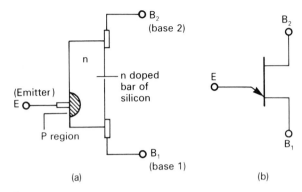

Fig. 9.5 (a) UJT construction; (b) BS symbol for a UJT

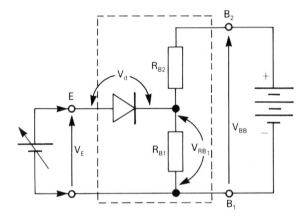

Fig. 9.6 UJT equivalent circuit

p-n junction now exists between the emitter and the silicon bar.

The operation of the device is best described by considering the equivalent circuit of the UJT (Fig. 9.6).

The emitter region is shown as a p-n diode and will exhibit the normal diode volt drop, e.g. $V_d \simeq 0.6$ V. The silicon bar is purely resistive with an interbase resistance (R_{BB}) of about 7 k. However since the emitter region is *off-centre* the values of R_{B1} and R_{B2} will be different, these values being determined by the actual position of the emitter.

Consider Fig. 9.6. With $V_E = 0$ V, B_2 is biased positive with respect to B_1 by the interbase voltage V_{BB}. The bar acts as a potential divider with the voltage V_{RB1} determined by the ratio of the two resistances R_{B1}, R_{B2}. This ratio

$$\frac{R_{B1}}{R_{B1} + R_{B2}}$$ is called the *intrinsic stand off ratio*

(η) and is quoted as a parameter in the manufacturer's specifications for the device. The voltage (V_{RB1}) that appears across R_{B1} is given by ηV_{BB}. If the emitter voltage (V_E) is now increased from 0 V there will come a point (V_{ep}) where the emitter junction becomes forward biased and current will flow from E to B_1. For this to happen V_E must overcome the junction potential (V_d) and the voltage across R_{B1} (ηV_{BB})

$$\therefore V_E = V_{ep} = V_d + \eta V_{BB}$$

since $\quad V_d \simeq 0.6 \quad V_{ep} \simeq \eta V_{BB} + 0.6$

When V_E reaches V_{ep} conduction occurs very rapidly with the result that the device exhibits a negative resistance characteristic as shown in Fig. 9.7.

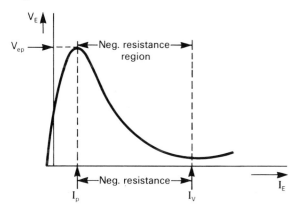

Fig. 9.7 UJT characteristic

Note The conduction is through the emitter to Base 1. The emitter-base 1 junction carries the main current.

UJT Specifications

There is no need for a particularly wide choice of unijunction transistor.

A common device is the 2N2646 UJT, which has the specifications shown in Table 9.1.

From an operating standpoint, the most important specification is η since this determines the emitter voltage required for conduction. Notice that this is quoted as a range together with a typical value, this means that individual transistors will perform differently despite the fact that they have the same code numbers. A unijunction transistor with η = 0.65 has an interbase voltage (V_{BB}) of 15 V. Calculate the emitter voltage required to make the device conduct:

$$V_E = V_{ep} = V_d + \eta V_{BB}$$
$$V_d \simeq 0.6$$
$$\eta = 0.65$$
$$V_{BB} = 15 \text{ V}$$
$$\therefore \quad V_{ep} = 0.6 + (0.65 \times 15) = 10.35 \text{ V}$$

Answer For conduction V_E must be 10.35 V.

The intrinsic stand off ratio for any individual UJT can be measured as shown in Practical Investigation 20.

Applications of the UJT

The very fast conduction characteristic of this transistor makes it particularly well suited to the generation of very sharp pulses. This is best illustrated by the relaxation oscillator circuit of Fig. 9.8.

At the instant of switch on, the voltage V_E will be zero and so the emitter junction of the UJT will be reverse biased and in a non-conducting state.

Capacitor C will charge via R and so V_E will increase exponentially following the law

$$V_C = V_E = V_{BB}(1 - e^{-t/CR}) \text{ Volts}$$

When V_E reaches the conduction point (V_{ep}) the UJT will switch on thus discharging C. The

Table 9.1 2N2646 data

Characteristic	Range	Typical values
η (intrinsic stand off ratio) when $V_{BB} = 10$ V	0.56–0.75	0.65
R_{BB} (interbase resistance) when $V_{BB} = 3$ V $I_E = 0$	4.7 kΩ–9.1 kΩ	7 kΩ
I_p (conduction point current) when $V_{BB} = 25$ V	5 μA (max)	0.4 μA
I_v (valley point current) when $V_{BB} = 20$ V	4 mA (min)	6 mA
V_{BB} max (max. interbase voltage)	35 V	

PRACTICAL INVESTIGATION 20

Unijunction Transistor Characteristics

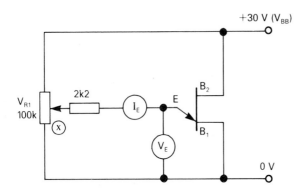

+30 V (V_{BB})

Equipment
Power supply
2N2646 UJT
100 k variable resistor
2 digital multimeters
2k2 resistor

Method
1   Build the circuit as shown above.
2   Set V_{R1} to its maximum value (point x).
3   Set V_{BB} to 30 V.
4   By reducing V_{R1} slowly, increase V_E while observing both I_E and V_E, you will notice the following:

  a) At the peak point (V_{ep}), V_E will start to fall.
  b) At the same point I_E will increase suddenly.

  Continue to reduce V_{R1} noting that I_E increases.

5   Reset V_{R1} to its maximum value (point x).
6   Repeat step 4 but this time increase V_E in steps until V_{ep} is reached recording V_E and I_E.

Results
1   Plot a graph of V_E against I_E and from this determine the conduction voltage (V_{ep}) of the UJT.
2   Determine the intrinsic stand off ratio (η) of your UJT using:

$$V_{ep} = V_d + \eta V_{BB} \therefore \eta = \frac{V_{ep} - V_d}{V_{BB}} \text{ (Assume } V_d \simeq 0.6 \text{ V)}$$

discharge current from C passes through R_b and will develop a voltage across it. This current will be of very short duration, consequently the output voltage waveform from B_1 will be a needle sharp pulse of short duration. These waveforms are shown in Fig. 9.9.

You will notice that the emitter voltage waveform is an approximate sawtooth caused by the charge and very fast discharge of C.

THOUGHT ⎯⎯⎯⎯⎯⎯⎯⎯⎯

Why is the output pulse from B_2 lower in amplitude than the output from B_1? The silicon bar is not split equally into R_{B1} and R_{B2} ∴ there will be a different voltage dropped across each section.

To develop an understanding of the pulse shape provided by the UJT it is a good idea to perform the practical investigation showing the UJT as a relaxation oscillator.

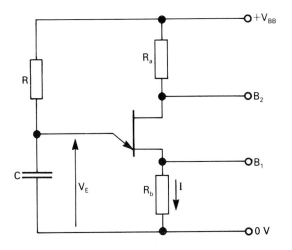

Fig. 9.8   The UJT relaxation oscillator

Emitter voltage
V_E

Output from
B_1

Output from
B_2

Fig. 9.9   Relaxation oscillator waveforms

The UJT as a triggering device

The type of pulses obtained from the output B_1 of the UJT are ideal for triggering thyristors and triacs. To achieve the control required it is necessary to use a CR network to determine the point at which the UJT conducts and then use the output pulse from B_1 to fire the control device. A circuit for achieving this is shown in Fig. 9.10.

Let us examine this circuit closely to see exactly how the control is accomplished. Notice that the thyristor is connected in series with the load so

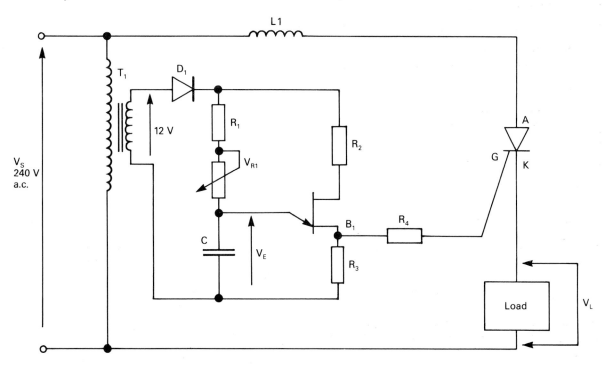

Fig. 9.10   Thyristor triggering using a UJT

PRACTICAL INVESTIGATION *21*

The UJT Relaxation Oscillator

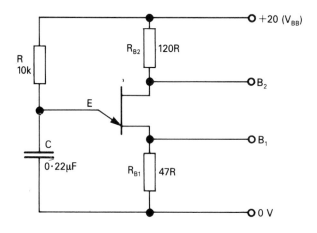

Equipment
UJT e.g. 2N2646
Oscilloscope (CRO)
47 R, 120 R, 10 k resistors
0.22 µF capacitor
Power supply

Method
1   Build the circuit as shown above.
2   Connect the power supply and adjust to give 20 V.
3   Using the CRO monitor V_E to check if the oscillator is working.
4   Sketch on a time-related scale waveforms at E, B_1 and B_2. In each case show: the d.c. level, amplitude, periodic time and duration of the pulses.

Results
1   The frequency of this oscillator can be determined theoretically using

$$f_o = \frac{1}{R_C \ln[1/(1 - \eta)]} \text{ Hz}$$

where η = intrinsic stand off ratio   ln = $\log_e$.
Using this equation together with the intrinsic stand off ratio value of your UJT calculate the theoretical output frequency and compare this to the actual output frequency.

that it switches the full mains voltage (remember L_1 is there to act as an interference suppressor).

The UJT is a low voltage device so a step-down transformer is used to provide a suitable operating voltage. The thyristor can only be switched on during the positive half cycles of the supply voltage so D_1 ensures that the unijunction circuit only receives a positive half cycle. V_{R1} and C form a CR network that will determine when, during the positive half cycle, the UJT will conduct.

When conduction occurs a sharp positive going pulse will appear on the gate of the thyristor and it will fire. Resistor R_4 is included to limit the gate current. Fig. 9.11 shows the waveforms associated with this circuit. Study them closely.

An interesting point to note is that the network V_{R1}, C determines the point in the positive half cycle that the UJT switches, and this produces the output pulse on B_1. However once the UJT conducts, C is discharged but may have time to

---

PRACTICAL INVESTIGATION 22

Thyristor Triggering Using a UJT

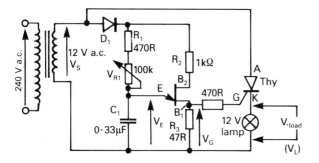

Equipment
Dual beam CRO
240:12 V transformer
12 V lamp (0.5 W max.)
Thyristor, e.g. T1C106
100 k variable resistor
Rectifier diode
0.33 μF capacitor
47 R, 2 × 470 R, 1 k resistors
UJT, e.g. 2N2646

Method
1 Build the circuit shown above.
2 Monitor V_s and V_L with the CRO and observe the effect of varying V_{R1}.
3 On a common time scale sketch the following waveforms: V_s, V_E, V_G and V_L for mid-point and maximum settings of V_{R1}.

Results
1 Compare the operating differences between this circuit and that of Investigation No. 19 and say why you think the UJT triggering circuit may be superior to the single phase control circuit.
2 Comment on the *firing* waveform (V_G).

---

charge and switch the UJT several times in a single half cycle! This does not matter at all since the thyristor will trigger on the first pulse and remain on until V_s passes through zero. The subsequent firing pulses will have no effect whatsoever. Practical Investigation 22 illustrates these points very well indeed.

A word of warning

Thyristors and triacs are normally used for the control of mains (240 V) operated loads, e.g. lamps and motors. The a.c. operation of these devices can be understood using a low voltage supply from a step-down transformer. On no account attempt to operate these circuits at a higher voltage than that stated in the investigations.

Self Assessment 8

Draw a circuit diagram that shows a UJT firing a triac that provides full wave control of a mains load. Refer to the last chapter for clues!

Gate Isolation Methods

You may have noticed that in a number of arrangements the gate trigger pulse to a power control device is supplied from a low voltage circuit. Thyristors and triacs are available in a variety of ratings and may be required to switch supply levels from low voltages up to several

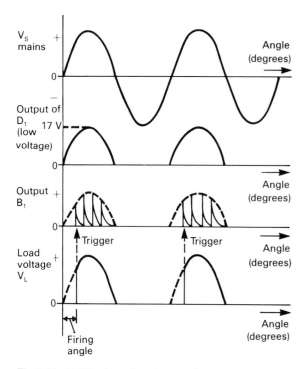

Fig. 9.11   UJT thyristor triggering waveforms

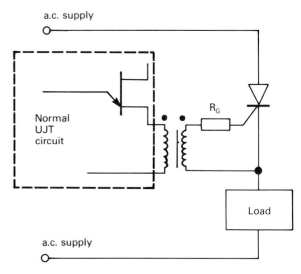

Fig. 9.12   Gate isolation using a pulse transformer

kilovolts. For safety reasons where high voltages are concerned, e.g. mains, there is a need to isolate the gate circuit from the supply voltage that is actually being switched. The reason for this is that during triggering the gate will become *live* to the supply peak voltage until the device is fully conducting. Consider a thyristor controlling a mains operated lamp. Mains voltage = 240 V a.c., peak voltage = 240 V × 1.414 ≃ 340 V.

If the thyristor receives a firing angle of 90° it will conduct when the supply is 340 V. As it conducts the gate will be at this potential until the device is fully turned 'on'. A voltage as low as 40 V may prove lethal so some form of isolation must be provided. Two common methods of providing isolation involve the use of either a pulse transformer or an opto-isolator.

The pulse transformer

The trigger pulse is applied to the primary of a transformer and the actual gate pulse is obtained from the secondary winding. In this way complete electrical isolation exists between the thyr-

istor or triac gate and the pulse circuit. The only connection between them is that provided by electromagnetic coupling. Fig. 9.12 shows a pulse transformer isolating the output from a UJT circuit triggering a thyristor. Note the *spot* convention that indicates the phase relationship between the primary and secondary windings.

The pulse transformer arrangement works equally well for triacs and if required, more than one secondary winding can be used to provide trigger pulses for a number of parallel devices.

The opto-isolator

Sometimes called opto-couplers these are available as a sealed dual-in-line (dil) package. This contains an infra-red, light emitting diode (LED) driving a photosensitive device. The input to the opto-isolator determines the output of the LED, the light from this controls the photosensitive component.

Electrical isolation up to 5 kV is achieved between input and output because the only connection between them is via a beam of light. A variety of opto-isolators are available, the type being determined by the photosensitive component that is employed. Common types are illustrated in Fig. 9.13 and show single isolators housed in a 6 pin package. Multiple isolators are available in 4 and 16 pin packages.

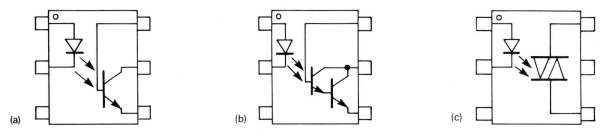

Fig. 9.13 (a) Opto-isolator using a single phototransistor; (b) Opto-isolator using a darlington pair phototransistor; (c) Opto-coupled triac

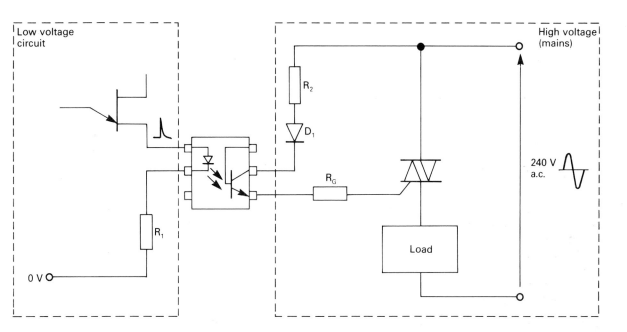

Fig. 9.14 Triac triggering using a single phototransistor opto-isolator

Typical applications for thyristor and triac triggering are shown in Figs 9.14 and 9.15.

Fig. 9.14 shows a UJT providing the normal triggering pulses; each pulse will cause the LED to flash and this in turn will forward bias the phototransistor thus causing a positive potential to appear on the gate, triggering the thyristor. Fig. 9.15 shows a similar circuit but here the LED in the opto-isolator switches on a phototriac which triggers the gate of the power triac thus controlling the load.

Burst firing

The circuits considered so far for the control of a.c. loads involve turning the control device 'on' part way through a half cycle. This phase control technique works very well but suffers from one major disadvantage. The triac or thyristor turns *on* and *off* during one cycle and although turn off occurs when the supply voltage passes through zero, turn on may well occur when the supply voltage is high, i.e. at its peak, see Fig. 9.16.

This switching of a high voltage generates

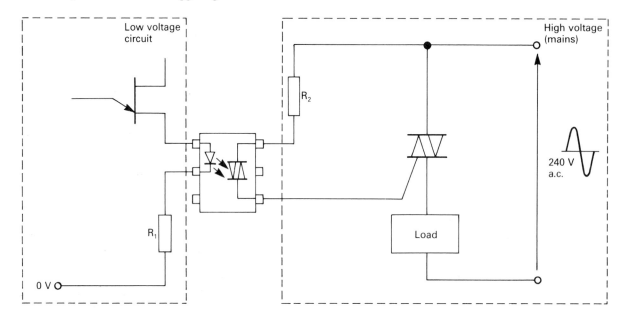

Fig. 9.15   Triac gate isolation using an optically coupled triac

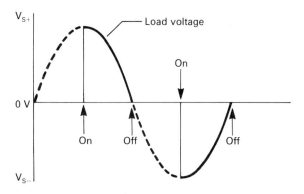

Fig. 9.16   Phase control switching using a triac

considerable electrical noise that requires suppression.

To avoid generation of such interference the device could be turned *on* and *off* when the supply voltage is zero. To achieve a.c. power control a technique called *burst firing*, or *integral cycling* may be used. This involves turning the control device 'on' for a specified number of complete cycles and then turning it 'off' for a number of cycles, as shown in Fig. 9.17.

Notice from the waveform diagrams that the triac is triggered with a firing angle of 0°. To achieve this a *zero voltage switch* is used to provide the gate trigger signal as the supply voltage crosses through the zero points. The triggering circuit is somewhat involved and beyond the scope of this book. However IC chips for burst firing controllers are available from most major suppliers, which simplifies design.

Power in the load is controlled by varying the ratio of cycles *ON* to cycles *OFF*. It follows that 1 cycle *ON* and 9 cycles *OFF* will give a low load

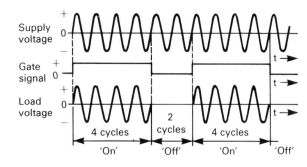

Fig. 9.17   'Burst firing' waveforms

power but 9 cycles *ON* and 1 cycle *OFF* will give high power.

This ratio of time *ON* to 'total time' is the controlling factor and is called the 'power transfer value' (*PV*).

$$PV = \left(\frac{\text{cycles on}}{\text{cycles on} + \text{cycles off}} \right)$$

To determine the actual power developed in the load it is only necessary to use Ohm's Law in conjunction with the power transfer value

$$V_{\text{Load}}(\text{rms}) = V_{\text{supply}}(\text{rms}) \times \sqrt{PV}$$
$$I_{\text{Load}}(\text{rms}) = I_{\text{supply}}(\text{rms}) \times \sqrt{PV}$$

$$\text{Load Power } (P_{\text{L}}) = \frac{V_{\text{supply}}^2}{R_{\text{Load}}} \times (PV)$$

Example

An integral cycling circuit uses a 240 V 50 Hz supply to control the power in a heating element with a resistance of 120 Ω.

Calculate the power developed in the load for the following firing arrangements:

(a)   20 cycles on 30 cycles off,
(b)   6 cycles on 4 cycles off,
(c)   5 cycles on 10 cycles off.

(a) $PV = \dfrac{\text{cycles on}}{\text{cycles on} + \text{cycles off}} = \dfrac{20}{20 + 30}$

$= \dfrac{20}{50} = 0.4$

$\text{Load power } (P_{\text{L}}) = \dfrac{V_{\text{supply}}^2}{R_{\text{Load}}} \times (PV)$

$= \dfrac{240^2}{120} (0.4) = 192 \text{ W}$

(b) $PV = \dfrac{6}{6 + 4} = \dfrac{6}{10} = 0.6$

$P_{\text{L}} = \dfrac{240^2}{120} (0.6) = 288 \text{ W}$

(c) $PV = \dfrac{5}{5 + 10} = \dfrac{5}{15} = 0.33$

$P_{\text{L}} = \dfrac{240^2}{120} (0.33) = 158 \text{ W}$

While this is a very useful method of providing interference-free control of power devices it has one drawback. The load power is literally controlled in *bursts*. Consequently this arrangement could not be used for lamp dimmers and motor speed controllers since pronounced flicker or jerky operation respectively would result.

It is, however, eminently suitable for loads that have a large *thermal time constant*, i.e. they have a slow reaction time to changes in voltage or current. Heating elements possess this characteristic and integral cycling is frequently used to control the temperature of soldering irons, ovens and electric furnaces.

Self Assessment 9

Burst firing is used to control the power in a 240 V, 9 A mains heater. If the circuit is adjusted to give 9 cycles on 3 cycles off calculate:

(a)   The load voltage
(b)   The load current
(c)   The load resistance
(d)   The load power.

Triggering Review

1   Thyristors and triacs are usually triggered by a gate *pulse*.
2   A Zener diode can be used to provide a gate *threshold* level for a thyristor.
3   A diac is a *triggerless triac* that will conduct in either direction once a breakdown voltage has been reached.
4   A unijunction transistor (UJT) relaxation oscillator will provide sharp pulses, ideal for gate triggering purposes.
5   The UJT provides an output pulse when its emitter voltage rises above a specified level given by:

$$V_{\text{E}} = V_{\text{ep}} = V_{\text{d}} + \eta V_{\text{BB}}$$

6 Where mains supply voltages are used the gate triggering circuit should always be electrically isolated from the gate of the control device.
7 Gate isolation can be achieved using a pulse transformer or an opto-isolator.
8 Loads with a large thermal time constant, e.g. heaters can be controlled using burst firing (integral cycling) methods.
9 The power in the load of a burst firing system is determined by the power transfer value PV given by:

$$PV = \left(\frac{\text{cycles on}}{\text{cycles on} + \text{cycles off}} \right)$$

$$\text{Load power } P_L = \left(\frac{V_{\text{supply}}^2}{R_{\text{Load}}} \right) \times PV \text{ Watts}$$

Self Assessment Answers

Self Assessment 8

For a triac to provide full wave control it must be fired during both positive and negative half cycles. To achieve this using a UJT means that a full wave rectifier must be used instead of the single diode of Fig. 9.10.

Self Assessment 9

(a) $V_{\text{Load}}(\text{rms}) = V_{\text{supply}}(\text{rms}) \sqrt{PV}$

$$PV = \left(\frac{\text{cycles on}}{\text{cycles on} + \text{cycles off}} \right) = \frac{9}{9+3}$$

$$= \frac{9}{12} = 0.75$$

$$V_{\text{Load}} = 240 \times \sqrt{0.75} = 207.8 \text{ V}$$

(b) $I_{\text{Load}}(\text{rms}) = I_{\text{supply}}(\text{rms}) \sqrt{PV} = 9 \times \sqrt{0.75} = 7.79 \text{ A}$

(c) $R_L = \dfrac{V_{\text{Load}}}{I_{\text{Load}}} = \dfrac{207.8}{7.79} = 26.7 \ \Omega$

(d) $P_L = \dfrac{V_L^2}{R_L} = \dfrac{(207.8)^2}{26.7} = 1.62 \text{ kW}$

or $P_L = V_L \times I_L = 207.8 \times 7.79 = 1.62 \text{ kW}$

or $P_L = I_L^2 \times R_L = (7.79)^2 \times 26.7 = 1.62 \text{ kW}$

or $P_L = \dfrac{V_{\text{supply}}^2}{R_L} \times (PV)$

$$= \frac{240^2}{26.7} \times (0.75) = 1.62 \text{ kW}.$$

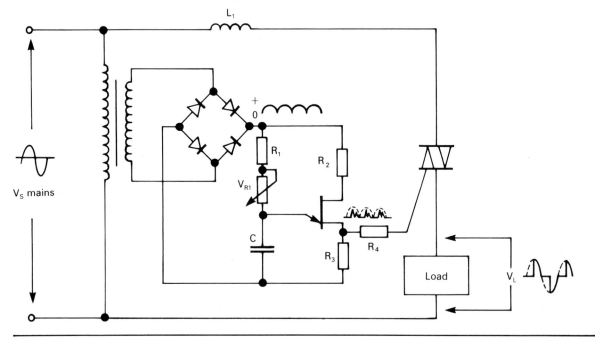

10
Fault Diagnosis

When any piece of electronic equipment fails to perform as expected it has 'failed'. The person who can repair it is often considered to be some sort of mystic or 'guru'. Many individuals who can successfully build circuits become powerless if they do not work properly. This is often most people's baptism into the world of electronics. There must be countless home electronics enthusiasts who have spent hours lovingly constructing an exciting project. The great moment comes, all is carefully checked: power supply, wiring and at last the *on* switch is pressed — nothing happens!! The intrepid constructor now feels helpless, a few tugs and wiggles of components here and there culminating in a grunt of dissatisfaction and disappointment as the project is cast aside to gather dust. If you have performed a number of the foregoing practical investigations you will have undoubtedly discovered that some circuits do not work *first time* for a variety of reasons. This scenario is acted out at all levels throughout the world of electronics, large companies probably have expensive equipment laying idle all for the sake of someone to repair it.

Today it is perfectly reasonable to expect all technician engineers to be able to fault diagnose circuits down to component level, effect the repair and then check the equipment, recalibrating it if necessary. It must be said at the very start that fault diagnosis is a logical process; the discovery of a failed component by random methods and guesswork is a very chancy business and will only yield results if

(a) there are very few components in the circuit,
(b) there is a history of failure, i.e. a particular television model may have an intrinsic fault

where a certain resistor burns out after a number of hours.

The repair service engineer arrives on the scene and listens gravely to the set of symptoms related by the owner. Immediately the back is off the set, there is a quick flurry with a soldering iron, the back is on again and all is well — almost like magic! The respected and almost holy engineer leaves the scene feeling ten feet tall. The reality is that he has seen the symptoms before and repaired countless identical sets.

The moral here is never guess the cause of a failure unless the guess is an educated one, based on past experience.

Component Failures

Now that the random method has been firmly put in its rightful place let us consider the realistic practical approaches that can be adopted. To do this we must take a brief look at how common components fail. Table 10.1 gives an indication of common component failures. This does not of course help if a circuit has been incorrectly wired or modified but it is of enormous value when trying to identify problem components.

Fault Finding Methods

Electronic systems may be considered as a series of blocks (Fig. 10.1). Most methods of fault diagnosis consist of tracing a signal through the

Table 10.1 Common component failures

Component	Common type of failure
Resistors	Open circuit or high in value *Note* resistors rarely fail by becoming *short-circuit*.
Variable resistors	Open circuit track *or* intermittent wiper contacts
Capacitors	Open circuit or short circuit.
Inductors	Open circuit winding, short circuit turns or short circuit between winding and core.
Transformers	Open circuit winding, short circuit turns or short circuit between winding and core/frame.
Diodes	Open or short circuit.
Transistors, FETs Thyristors Triacs	Open or short circuit at any junction, i.e. between electrodes.

blocks until it either disappears or distorts. Before the search commences, however, there is a preliminary sequence that is so obvious it is often ignored. The vague 'it doesn't work' is no basis from which to start a logical search! Test the equipment personally and identify the nature of the fault. Then proceed in the following manner:

1   Use your senses. Upon opening up the equipment look for signs of damage, e.g. broken wires, boards etc. Indications of burning or overheating can usually be both seen and smelt — but remember an overheated component indicates the area of the fault, it has probably suffered excess current flow because something else has failed!
2   If there is no life in the equipment at all, check the mains fuse, plug and lead first. *A great many items have been totally dismantled, only to find that a wire had become disconnected in the mains plug!*
3   If there is power but a malfunction is suspected, run through a procedure that will allow the operation to be thoroughly tested so that symptoms can be noted. For a piece of equipment like a signal generator this will consist of observing the output on all ranges with an oscilloscope and checking for distortion. Where a complicated electronic instrument or system is concerned, most manufacturers supply test procedure information with the user manual. As a result of this functional check the symptoms will have been accurately noted and the quest for the faulty component can now begin in earnest.

Input to output/output to input method

The signal route through a system has been identified and the circuit itself broken down into a series of stages or blocks shown in Fig. 10.1. This method consists of applying an input signal and then checking for a signal at the output of each stage from 1 to 6. When the signal disappears the faulty block or stage has been identified. An alternative method is to monitor the output and inject a signal sequentially into each block moving towards the input from 6 to 1. When the output disappears, the faulty stage has been discovered.

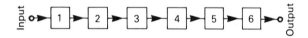

Fig. 10.1   An electronic system as a series of blocks

These two methods are logical and will produce results but can be quite lengthy if a large number of blocks are involved.

The half-split method

Where there are many stages, a quick method is as follows:

1   Apply a signal to the input.
2   Check for a signal approximately half-way through the system, e.g. at the output of block (3) in Fig. 10.1.

3 If a signal is present, blocks 1 to 3 are ok so the fault is between 4 and 6.
4 Check half-way through the remainder, e.g. block 5, if ok the fault is in block 6.
5 If no output at 5 check at 4, i.e. the half-way point again.

This is a very logical and economical testing method since it reduces the number of measurements required. Whichever method is concerned you must remember that real life tends to upset the theoretical ideas. Complications will invariably occur for the following reasons:

1 It may not be physically possible to make measurements at the desired points in a circuit.
2 Circuits may consist of a number of stages that converge at a single point, resulting in the input to a stage consisting of outputs from a number of other stages as shown in Fig. 10.2.
3 Divergence may occur where the output of one stage splits to become the input to a number of stages (Fig. 10.3).
4 Feedback may exist where a signal from one stage in a system may be fed back to an earlier stage (Fig. 10.4).

These complications may occur individually or collectively in a complete system. Accurate pinpointing of a faulty stage will involve careful

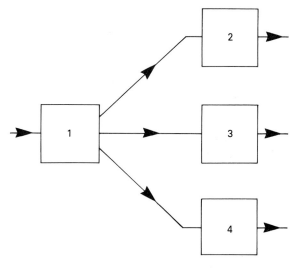

Fig. 10.3 Divergence

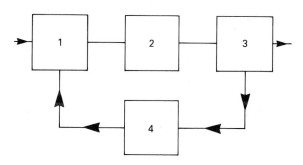

Fig. 10.4 Feedback

interpretation of measured results if the task is to be accomplished quickly.

Identifying the Faulty Component

Once the problem stage has been found diagnosis must be used to find the 'rogue' component. To do this certain *test instruments* will be required. The basic set of equipment will consist of the usual list: power supply, signal generator, voltmeter and oscilloscope. The majority of tests will be carried out using the voltmeter to measure d.c. levels and the CRO to trace a.c. signal paths.

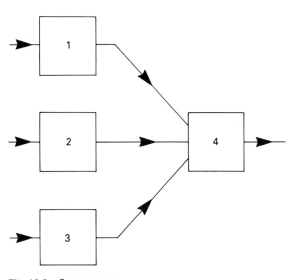

Fig. 10.2 Convergence

THOUGHT

Why bother measuring d.c. levels, surely it is sufficient just to trace the signal?
It is true that this will tell us a lot, but we must identify the single component that is faulty and often only d.c. levels will indicate this.

Measuring d.c. levels

Consider the circuit of Fig. 10.5.

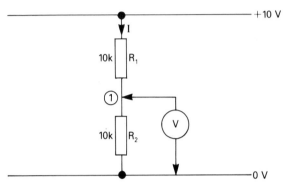

Fig. 10.5   Measuring d.c. levels

The voltage measured between 0 V and point 1 can be estimated using Ohm's Law as follows:

$$I = \frac{V_{dc}}{R_1 + R_2} \text{ and since } V = \text{voltage across } R_2$$

$$V = R_2 \times I$$

This can be expressed using the single equation

$$V = \frac{V_{dc}}{R_1 + R_2} \times R_2$$

We would expect under normal conditions the voltage at 1 to be 5.0 V, this would indicate that both R_1 and R_2 were ok! *Supposing the voltage was measured and found to be 0 V, what would be the fault?* Clearly since the current path is a series one we would assume R_1 has failed open circuit. *What if the measured value were found to be 10 V?*

CAREFUL   Do not fall into the trap of assuming R_1 has failed short-circuit! We know this virtually never happens! Think on — Ohm's Law tells us

that the voltage dropped across a resistor is directly proportional to the current flowing through it. Therefore if the current flowing through R_1 is zero, the voltage dropped across it will be zero so consequently point 1 will be at the same potential as the d.c. supply (V_{dc}).

THOUGHT

So both ends of R_1 will be at the same potential? — Yes, but this assumes the voltmeter itself will draw no current. In practice some current must flow through the meter (otherwise it would not indicate), if it is a good quality instrument this will be a very low current and the voltage at point 1 will be almost V_{dc}.

You can probably already see that measurements can only be interpreted using an understanding of circuit concepts and behaviour, try the following example.

Self Assessment 10

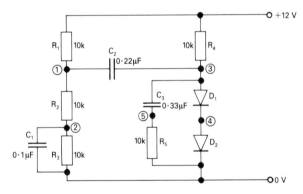

1   Study the above circuit and estimate the voltages at the points indicated (all voltages considered with respect to the 0 V line.)
2   Comment on the effect that the capacitors will have on the d.c. levels in this circuit.

Voltmeter Loading

Many manufacturers' circuit diagrams include normal d.c. voltage levels at important points so that the service engineer will know the expected correct voltage level, he can now compare voltage measurements made on the faulty equipment with these and make a value judgement. It is of considerable importance that the characteristics of the voltmeter are understood in order to appreciate the effects it will have on a circuit.

Every voltmeter has some resistance: when making a measurement therefore, it should be considered as a resistor that is connected in parallel with the component across which the voltage is being measured (Fig. 10.6).

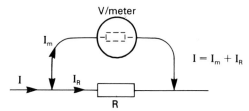

Fig. 10.6 Voltmeter resistance

The current is going to split. Some will flow through the circuit (I_R), and some through the meter (I_m). This means that whenever a voltmeter is used on a circuit the circuit will be disturbed. The extent to which it is disturbed will be determined by the resistance of the meter (R_m). The higher the voltmeter resistance the better! It will then cause least disturbance to the circuit and thus yield more accurate results. Study the circuit of Fig. 10.7.

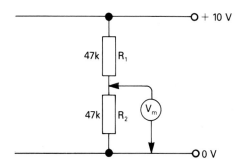

Fig. 10.7 Voltmeter loading

The theoretical voltage that should be indicated is of course 5.0 V. Now supposing voltmeter (V_m) has a resistance R_m of 20 kΩ? The circuit can be re-drawn as Fig. 10.8.

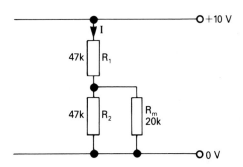

Fig. 10.8 Equivalent circuit of a voltmeter

R_m is in parallel with R_2 and R_1 is in series with this parallel network.

$$R_T = R_1 + R_2 \| R_m$$

(remember! $\|$ means in parallel with)

$$R_2 \| R_m = \frac{47\,\text{k} \times 20\,\text{k}}{47\,\text{k} + 20\,\text{k}} = 14.03\,\text{k}\Omega$$

$$R_T = 47\,\text{k} + 14.03\,\text{k} = 61.03\,\text{k}\Omega.$$

This will give a current (I) of:

$$I = \frac{10}{61.03 \times 10^3} = 164\,\mu\text{A}$$

The voltmeter will indicate the voltage across the parallel network $R_2 \| R_m$

$$V_m = I \times 14.03\,\text{k}$$
$$= 164 \times 10^{-6} \times 14.03 \times 10^3 = 2.3\,\text{V}$$

So using a voltmeter with an internal resistance of 20 k will produce a reading of 2.3 V where 5.0 V is expected. *BUT* there is nothing wrong with the circuit, the voltmeter is simply loading it! This shows, I hope, how careful you must be when selecting a voltmeter for making circuit measurements.

Digital meters

Most digital meters have a fixed input resistance that is specified on the instrument, usually 1 MΩ. This means that whatever d.c. range is selected the instrument will act like a 1 MΩ resistor when it is introduced into the circuit.

Analogue meters

The analogue voltmeter or multimeter is really a micro-ammeter calibrated to display voltage. A current must flow through the meter coils in order to move the pointer. The lower the current required to produce full scale deflection (fsd) the more sensitive the meter is considered to be.

Example

A meter that requires only 500 μA for full scale deflection is more sensitive than one that requires 1 mA for fsd.

From this it follows that the coils of the meter must possess resistance (R_m) and it is this resistance that determines how many volts are required across the coils to produce the current required for fsd.

Most modern instruments have meter movements that require a current (I_m) of 50 μA for fsd and have a resistance (R_m) of 20 kΩ. This means that the voltage required for fsd = $I_m \times R_m$ = $50 \times 10^{-6} \times 20 \times 10^3 = 1.0$ V.

∴ a voltmeter with a 50 μA movement will measure 1 V full scale.

Now most voltmeters are multi-range, which means that to measure voltages greater than 1.0 V, resistors are added in series with the meter coils. Changing ranges selects different resistors. This means the meter's internal resistance will depend upon the range that is selected. On all ranges, however, the meter resistance is ultimately governed by the movement itself and if this is 50 μA, 1 volt is required for fsd. This means that the meter sensitivity can be expressed as 20 kΩ/V. Since the movement resistance is given by

$$R_m = \frac{V_m}{I_m} = \frac{1}{50 \, \mu A} = 20 \, k\Omega$$

Therefore on the 3 V range the meter resistance will be 3 × 20 k = 60 kΩ and on the 10 V range it will be 10 × 20 k = 200 kΩ. So the awful fact is: *a multirange analogue meter has the highest resistance on its highest range.*

THOUGHT

So if the instrument is set to the 10 V range and the range is lowered, the meter resistance falls? This is true and as its resistance becomes lower it draws more current from the circuit it is measuring and produces false readings.

Example

You are using a meter on the 10 V range and the indicated voltage appears to be 0.9 V. You change to the 1.0 V range hoping for a more accurate reading only to discover that the meter indicates 0.4 V. The meter is not faulty it simply has a lower resistance and is drawing more current away from the circuit.

The golden rule is always use a voltmeter that has the highest resistance.

Practical Investigation 23 highlights this peculiarity.

From this investigation you will have discovered that the most accurate voltmeter will be the one that has the highest resistance since it will only require a small amount of current to operate and thus offer minimum disturbance to the circuit. However, you should have made another perhaps more significant finding. This is that the voltmeter's resistance should be high when compared to the resistance across which the voltage is being measured, i.e. a 20 kΩ/V meter on the 1 volt range has a resistance of 20 kΩ, if it is measuring the voltage across a 2 Ω resistor it will be quite accurate. If it is used to measure the voltage across a 100 MΩ resistor it will be quite inaccurate.

Interpretation of voltmeter readings

It is very simple to make voltage measurements at a number of points in a circuit just as it is easy for a doctor to carry out a series of tests on a patient, but, once the results are obtained, they must be carefully analysed and a judgement made as to their significance. To do this we must have a 'norm' or reference — a doctor can judge a patient's temperature or blood pressure to be high because he has an idea of what the *normal* level is. In fault diagnosis we need to know approximately what a voltage should be in order

PRACTICAL INVESTIGATION *23*

Voltmeter Loading

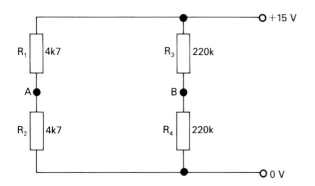

Equipment
Power supply
Digital voltmeter
10 V analogue voltmeter
(20 kΩ/V)
2 off 4k7 resistors
2 off 220 k resistors
Breadboard

Method
1   Construct the circuit as shown.
2   Measure and record the voltage between the 0 V line and points A and B using

   (a)   the digital voltmeter,
   (b)   the analogue voltmeter.

Results
1   From your results deduce which instrument is the most accurate and state why.
2   Under what circumstances may both a low resistance and a high resistance voltmeter yield the same accurate results?

to decide if our measured voltage is in any way abnormal. For this reason most equipment manufacturers supply circuit diagrams that have test points indicated together with the typical voltage value printed by the side. Such a diagram provides a set of *normal* voltage readings, you will be able to compare any fault condition readings with these and make a diagnosis as to the faulty component. This is not easy, it requires knowledge, understanding and an ability to think logically. Practical Investigation 24 requires that you build a simple amplifier, take *normal* voltage readings and then introduce fault conditions that allow you to develop an understanding of the way a voltmeter will indicate the types of fault that can occur.

If you performed this investigation you will have discovered that some types of fault can lead to surprising readings. It is difficult and unwise to use 'hard and fast' rules when di-

agnosing faults from voltage measurements. There are an infinite variety of circuits available in the world to go 'wrong' and the skilled technician should be able to trace a fault down to component level in any circuit. Experience and practice as always yield the best results, but remember also that 'dogma' is the curse of the engineer — never convince yourself that you know what is wrong with a circuit before you have checked the operation for yourself. Every engineer at some time has completely dismantled a piece of apparatus to repair the 'fault' he knew existed only to find, after considerable time, that the fault was a missing fuse, a wire adrift in the plug or most humiliating of all the apparatus was just switched off — this sobering experience should only happen once. It is a learning experience. Fault diagnosis, like electronics, generally must be approached with enthusiasm and an open mind, working from the

Self Assessment 11

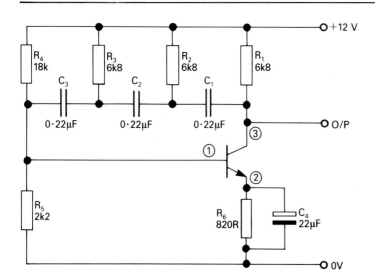

Test Points	①	②	③
Normal reading	1·3 V	0·7 V	6·25 V

The above diagram shows a phase shift oscillator circuit together with a set of normal voltage readings. The measurements were taken with respect to the 0 V line using a 20 kΩ/V voltmeter. For each of the given fault conditions, indicate the faulty component and the nature of the fault giving reasons supporting your answer.

Test points	1	2	3
Fault 1	0.00	0.00	11.98
Fault 2	2.40	2.70	1.75
Fault 3	1.30	0.80	11.98
Fault 4	0.75	0.00	0.02
Fault 5	1.75	1.10	1.75

known towards the unknown in a logical and thoughtful manner.

Fault Diagnosis Review

1   Fault diagnosis must be performed in a logical step-by-step fashion.
2   The random method of fault diagnosis is really guesswork and should only be used if there is a known *history* of failure in a circuit.
3   Typical fault tracing methods include the input-to-output, output-to-input and half-split methods.
4   All methods will be complicated by the exist-

ence of feedback loops, convergence and divergence.
5   Components fail in common or typical ways but resistors seldom fail short-circuit.
6   A voltmeter can be used to identify a faulty component by measuring the d.c. levels.
7   All voltmeters possess resistance and since they are connected in parallel will disturb or load a circuit, by drawing current.
8   The most accurate voltmeter is one that has a very high internal resistance when compared to the resistance of the component across which it is connected.
9   Voltmeter readings have to be interpreted by comparing the measured value with a known *normal* d.c. level. This is often provided on circuit diagrams by equipment manufacturers.

PRACTICAL INVESTIGATION **24**

Component Fault Diagnosis

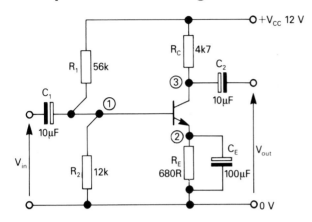

Equipment
Power supply
Signal generator
Oscilloscope
Small signal transistor, e.g. BC108
$2 \times 10\ \mu$F capacitors, 100 μF capacitor
680 R 4k7, 12 k, 56 k resistors
Breadboard

Method
1   Build the circuit shown above.
2   Connect a signal generator monitor V_{in} and V_{out} and check that the circuit is working.
3   Disconnect the signal generator — the circuit is now operating under quiescent conditions.
4   Using the voltmeter measure the voltages between the 0 V line and test points (1), (2) and (3) and enter the results in the table below.
5   One at a time, simulate each of the fault conditions listed in the table and then remeasure the voltage at each point entering the results in the table.

Test points	1	2	3
Normal reading			
R_1 o/c (open circuit)			
R_2 o/c			
R_E o/c			
R_C o/c			
C_E o/c			
C_E s/c (short circuit)			
Emitter-base Junction o/c			
Emitter-base Junction s/c			
Base-collector Junction o/c			
Base-collector Junction s/c			
Collector-emitter Junction o/c			
Collector-emitter Junction s/c			

Self Assessment Answers

Self Assessment 10

1 *Voltages at points (1) and (2)*
Using Ohm's Law R_1, R_2 and R_3 all have the same value, $\frac{1}{3}$ of the supply voltage is dropped across each resistor.
Voltage at point (1) = 8 V.
Voltage at point (2) = 4 V.

Voltage at points (3) and (4)
D_1 and D_2 are diodes each having a volt drop of approximately 0.6 V.
Voltage at point (3) = 1.2 V.
Voltage at point (4) = 0.6 V.

Voltage at point (5)
C_3 is open circuit to d.c. $\therefore$ voltage at point (5) = 0 V.

2 Capacitors can be considered to be open circuit (i.e. infinite resistance) to d.c. and will therefore have no effect on d.c. levels in a circuit.

Self Assessment 11

Fault 1   0 V on point 1 indicates no base bias to the transistor. This will mean the transistor is off so point 3 will float up to the supply.
Deduction R_4 o/c

Fault 2   2.4 V on point 1 indicates the potential divider is not working, resulting in a higher base bias. Consequently the transistor is saturated giving a low voltage at test point 3. R_5 o/c

Fault 3   Base bias voltage point 1 correct *but* the higher reading on point 3 indicates no current is flowing. The low reading on point 2 shows there is not the usual 0.6 volt drop across the base-emitter junction but this is because there is only the voltmeter current flowing. $\therefore$ R_6 o/c

Fault 4   Low base bias voltage together with very low voltage at point 3 indicates saturation has occurred. The 0 V on point 2 indicates C_4 s/c.

Fault 5   Voltages on points 1 and 2 show that the base-emitter junction is ok but the fact that points 1 and 3 are the same indicates a *base-collector s/c*.

11

Data Sheets

Transistors

type	material	case	application	P_T	I_C	V_{CEO}	V_{CBO}	h_{FE}	f_T (typ)
AC127	NPN Ge	TO1	Audio output	340 mW	500 mA	12 V	32 V	50	2·5 MHz
AC128	PNP Ge	TO1	Audio output	700 mW	−1 A	−16 V	−32 V	60-175	1·5 MHz
AD149	PNP Ge	TO3(A)	Audio output	*22·5 W at 50°C	−3·5 A	−50 V	−50 V	30-100	0·5 MHz
AD161 ⎫ Pair	NPN Ge	SO55(A)	Audio matched pair	*4 W at 72°C	3 A	20 V	32 V	50-300	3 MHz
AD162 ⎭	PNP Ge			*6 W at 63°C	−3 A	−20 V	−32 V	50-300	1·5 MHz
AF127	PNP Ge	TO72(A)	I.F. Applications	60 mW	−10 mA	−20 V	−20 V	150 ◆	75 MHz
BC107	NPN Si	TO18	Audio driver stages (complement BC177)	360 mW	100 mA	45 V	50 V	110-450	250 MHz
BC108	NPN Si	TO18	General purpose (complement BC178)	360 mW	100 mA	20 V	30 V	110-800	250 MHz
BC109	NPN Si	TO18	Low noise audio (complement BC179)	360 mW	100 mA	20 V	30 V	200-800	250 MHz
BC142	NPN Si	TO39	Audio driver	800 mW	800 mA	60 V	80 V	20 (min.)	80 MHz
BC143	PNP Si	TO39	Audio driver	800 mW	−800 mA	−60 V	−60 V	25 (min.)	160 MHz
BC177	PNP Si	TO18	Audio driver stages (complement BC107)	300 mW	−100 mA	−45 V	−50 V	125-500	200 MHz
BC178	PNP Si	TO18	General purpose (complement BC108)	300 mW	−100 mA	−25 V	−30 V	125-500	200 MHz
BC 179	PNP Si	TO18	Low Noise Audio (complement BC109)	300 mW	−100 mA	−20 V	−25 V	240-500	200 MHz
BC182L	NPN Si	TO92(A)	General purpose	300 mW	200 mA	50 V	60 V	100-480	150 MHz
BC183L	NPN Si	TO92(A)	General purpose (complement BC213L)	300 mW	200 mA	30 V	45 V	100-850	280 MHz
BC184L	NPN Si	TO92(A)	General purpose	300 mW	200 mA	30 V	45 V	250 (min.)	150 MHz
BC212L	PNP Si	TO92(A)	General purpose	300 mW	−200 mA	−50 V	−60 V	60-300	200 MHz
BC213L	PNP Si	TO92(A)	General purpose (complement BC183L)	300 mW	−200 mA	−30 V	−45 V	80-400	350 MHz

Type	Polarity	Case	Application	Power	Current	V1	V2	hFE	Frequency
BC214L	PNP Si	TO92(A)	General purpose	300 mW	-200 mA	-30 V	-45 V	140-600	200 MHz
BC237B	NPN Si	TO92(B)	Amplifier	350 mW	100 mA	45 V	—	120-800	100 MHz
BC307B	PNP Si	TO92(B)	Amplifier	350 mW	100 mA	45 V	50 V	120-800	280 MHz
BC327	PNP Si	TO92(B)	Driver	625 mW	-500 mA	-45 V	-50 V	100-600	260 MHz
BC337	NPN Si	TO92(B)	Audio driver	625 mW at 45 °C	500 mA	45 V	50 V	100-600	200 MHz
BC441	NPN Si	TO39	General purpose (complement BC461)	1 W	2 A peak	60 V	75 V	40-250	50 MHz (min.)
BC461	PNP Si	TO39	General purpose (complement BC441)	1 W	-2 A peak	-60 V	-75 V	40-250	50 MHz
BC477	PNP Si	TO18	Audio driver stages	360 mW	-150 mA	-80 V	-80 V	110-950	150 MHz
BC478	PNP Si	TO18	General purpose	360 mW	-150 mA	-40 V	-40 V	110-800	150 MHz
BC479	PNP Si	TO18	Low noise audio amp.	360 mW	-150 mA	-40 V	-40 V	110-800	150 MHz
BCY70	PNP Si	TO18	General purpose	360 mW	-200 mA	-40 V	-50 V	150	200 MHz
BCY71	PNP Si	TO18	Low noise general purpose	360 mW	-200 mA	-45 V	-45 V	100-400	200 MHz
BD131	NPN Si	TO126^m	General purpose – medium power	*15 W at 60 °C	3 A	45 V	70 V	20 (min.)	60 MHz
BD132	PNP Si	TO126^m	General purpose – medium power	*15 W at 60 °C	-3 A	-45 V	-45 V	20 (min.)	60 MHz
BD131 / BD132 (Pair)	NPN Si / PNP Si	TO126^m	Audio matched pair	*15 W at 60 °C / *15 W at 60 °C	3 A / -3 A	45 V / -45 V	70 V / -45 V	20 (min.) / 20 (min.)	60 MHz / 60 MHz
BD135	NPN Si	SOT32^m	Audio driver	*12·5 W at 25 °C	1·5 A	45 V	45 V	40-250	50 MHz
BD136	PNP Si	SOT32^m	Audio driver	*12·5 W at 25 °C	-1·5 A	-45 V	-45 V	40-250	75 MHz
BD437 / BD433	NPN Si / PNP Si	TO126^m	Power switching complementary	*36 W at 25 °C	4 A	45 V	45 V	40	3 MHz

Power Transistors

Ic(AV) max Amps	Ptot watts @25°C	NPN PNP	Package and Pin Connection	VCEO(V)					
				18—36	40—50	60	80	100	120—150
0.15	1.5	NPN	TO-39	BFW16A (25V) ●					
0.35	7	NPN	TO-39		2N3553 (40V) ●				
0.4	3.5	NPN	TO-39	2N4427 (20V) ●					
0.4	5	NPN	TO-39	2N3866 (30V) ●					
0.7	5	NPN	TO-39		2N3053 (40V)				
0.7	5	NPN	TO-39		2N3053-NSC (40V)				
1	4	NPN	X-03	AD161 (20V) ■					
1	6	PNP	X-03	AD162 (20V) ■					
1	8	NPN	TO-126		BD135 (45V)	BD137	BD139		
1	8	PNP	TO-126		BD136 (45V)	BD138	BD140		
1	10**	NPN	SOT-48/3	BLX68 (18V) ●					
1	30	NPN	TO-220		TIP29 (40V)	TIP29A	TIP29B	TIP29C	
1	30	PNP	TO-220		TIP30 (40V)	TIP30A	TIP30B	TIP30C	
1.5	12.5	NPN	TO-126				BD230		
1.5	12.5	PNP	TO-126				BD231		
2	25	NPN	TO-126		BD233 (45V)	BD235	BD237		
2	25	PNP	TO-126		BD234 (45V)	BD236	BD238		
3	15	NPN	TO-126		BD131 (45V)				
3	15	PNP	TO-126		BD132 (45V)				
3	25	NPN	TO-66						2N3441 (140V)
3	30	NPN	TO-126		2N4918 (40V)	2N4919	2N4920		
3	30	PNP	TO-126		2N4921 (40V)	2N4922	2N4923		
3	30	NPN	TO-220			BD239A		BD239C	
3	40	PNP	TO-220					BD240C	
3	40	NPN	TO-220		TIP31 (40V)	TIP31A	TIP31B	TIP31C	
3	40	PNP	TO-220		TIP32 (40V)	TIP32A		TIP32C	
3	40	NPN	TO-220			TIP31A-TI		TIP31C-TI	
3	40	PNP	TO-220			TIP32A-TI			
3	62	NPN	SOT-56	BLY93A (36V) ●					
4	25	NPN	TO-66			2N3054			
4	36	NPN	TO-126	BD433 (22V)					
4	36	PNP	TO-126	BD434 (22V)					
4	36	NPN	TO-126	BD435 (32V)					
4	36	PNP	TO-126	BD436 (32V)					
4	40	NPN	TO-126		BD437 (45V)	BD535			
4	40	PNP	TO-126		BD438 (45V)	BD536			
4	40	NPN	TO-220		2N5296 (50V)				
4	40	NPN	TO-126		MJE521 (40V)				
4	40	PNP	TO-220		MJE371 (40V)				
4	40	NPN	TO-126		2N5190 (40V)	2N5191	2N5192		
4	40	PNP	TO-126		2N5193 (40V)	2N5194	2N5195		
4	40	NPN	TO-220				2N6123		

Group	W	Type	Case					
5	5	NPN	TO-39		BFX34			
	15	NPN	TO-126		BDX35			
	15	NPN	TO-126		BDX36			
	40	NPN	TO-220	BD947 (45V)	BD241A	BDX37	BD241C	BD955 (120V)
	40	PNP	TO-220	BD948 (45V)			BD954	BD956 (120V)
6	10	NPN	TO-126	BUP40 (50V)				
	10	PNP	TO-126	BUP41 (50V)				
	65	NPN	TO-220	TIP41 (40V)	TIP41A	TIP41B	TIP41C	
	65	NPN	TO-220					
	65	PNP	TO-220	TIP42 (40V)	TIP42A, TIP42A-TI		TIP42C	
7	40	PNP	TO-220		2N6107 (70V)			2N6476 (120V)
	40	PNP	TO-220					
	40	NPN	TO-220			2N6292		2N6474 (120V)
	40	NPN	TO-220			2N5496		
	50	NPN	TO-220	2N6109 (50V)				
8	60	NPN	TO-220	BD201 (45V)	BD203	BDX77	BD243C	
	60	PNP	TO-220	BD202 (45V)	BD204	BDX78	BD244C	
	65	NPN	TO-220		BD243A		BDX95	
	65	PNP	TO-220		BD244A		BDX96	
	90	NPN	TO-3		BDX92	BDX94		
	90	PNP	TO-3					
10	40	NPN	TO-3		BDY92	BDY91	BDY90	
	75	PNP	TO-220		2N6099	2N6101 (70V)		
	75	PNP	TO-220		MJE2955T			
	80	NPN	SOT-93		MJE3055T			
	80	NPN	SOT-93		TIP33A			
	90	PNP	TO-220		TIP34A			
	90	NPN	TO-220					
	117	NPN	TO-3					
	150	NPN	TO-3		2N3791	2N3716	BDT95	2N3442 (140V)
	150	PNP	TO-3			2N3792	BDT96	
	150	PNP	TO-3			2N5876		
	150	NPN	TO-3			2N5878		

' @90°C   '' @70°C

■ Germanium   ● RF type

Power Transistors (continued)

Ic (AV) max Amps	Ptot watts @ 25°C	NPN PNP	Package and Pin Connection	VCEO (V)					
				18—36	40—50	60	80	100	120—150
12	40	NPN	TO-3						BUV27A (150V)
	65	NPN	TO-220						
15	90	PNP	SOT-93				TIP2955 (70V)		
	90	PNP	SOT-93				TIP3055 (70V)		
	90	NPN	TO-220					BD743C	
	90	PNP	TO-220					BD744C	
	115	NPN	TO-3			2N3055H			
	115	NPN	TO-3			2N3055			
	125	PNP	TO-220			BDT81			
	125	PNP	TO-220			BDT82			
	150	NPN	TO-3					BDT85	MJ15015 (120V)
	180	NPN	TO-3				2N6254		
20	140	NPN	TO-3					2N5038 (90V)	MJ15003 (140V)
	250	NPN	TO-3				2N5039 (75V)		MJ15004 (140V)
	250	PNP	TO-3						
25	90	NPN	SOT-93			TIP35A		TIP35C	
	90	PNP	SOT-93			TIP36A		TIP36C	
	200	PNP	TO-3			2N5883	2N5884		
	200	NPN	TO-3			2N5885	2N5886		
30	140	NPN	TO-3						2N5672 (120V)
	150	NPN	TO-3		2N3771 (40V)	2N3772			2N3773 (140V)
	150	NPN	SOT-93			BUW48			
	150	NPN	TO-3				BUW49		
	200	NPN	TO-3		2N5301 (40V)	2N5302		MJ802 (90V)	
40	200	NPN	TO-3					2N6328	
	200	PNP	TO-3					2N6331	
50	250	NPN	TO-3						BUT90 (125V)
	250	NPN	TO-3						BUV60 (125V)
	250	NPN	TO-3						BUV20 (125V)
	150	NPN	SOT-93					BUP44	
60	150	NPN	SOT-93			BUP43			
90	300	NPN	TO-3				BUP49		
100	300	NPN	TO-3			BUP48			

Transistor Pin Connections

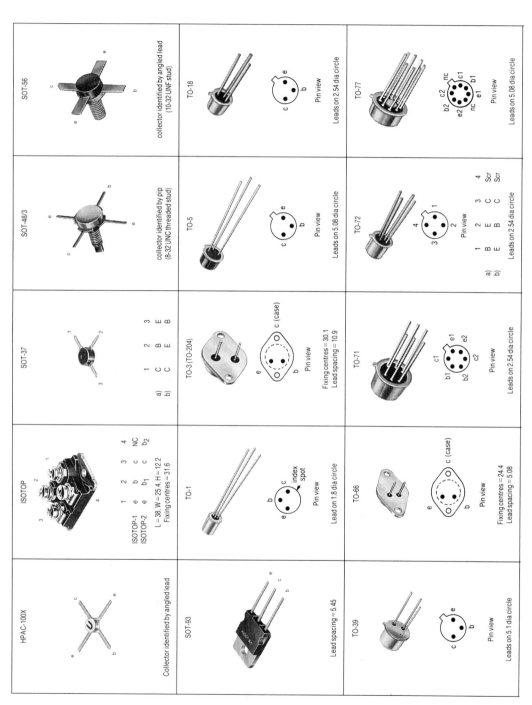

HPAC-100X

Collector identified by angled lead

ISOTOP

	1	2	3	4
ISOTOP-1	e	b	c	NC
ISOTOP-2	e	b₁	c	b₂

L = 38, W = 25.4, H = 12.2
Fixing centres = 31.6

SOT-37

	1	2	3
a)	C	B	E
b)	C	E	B

SOT-48/3

collector identified by pip
(8-32 UNC threaded stud)

SOT-56

collector identified by angled lead
(10-32 UNF stud)

SOT-93

Lead spacing = 5.45

TO-1

Pin view

Lead on 1.8 dia circle

TO-3 (TO-204)

c (case)

Pin view

Fixing centres = 30.1
Lead spacing = 10.9

TO-5

Pin view

Leads on 5.08 dia circle

TO-18

Pin view

Leads on 2.54 dia circle

TO-39

Pin view

Leads on 5.1 dia circle

TO-66

c (case)

Pin view

Fixing centres = 24.4
Lead spacing = 5.08

TO-71

Pin view

Leads on 2.54 dia circle

TO-72

	1	2	3	4
a)	B	E	B	Scr
b)	E	B	C	Scr

Pin view

Leads on 2.54 dia circle

TO-77

Pin view

Leads on 5.08 dia circle

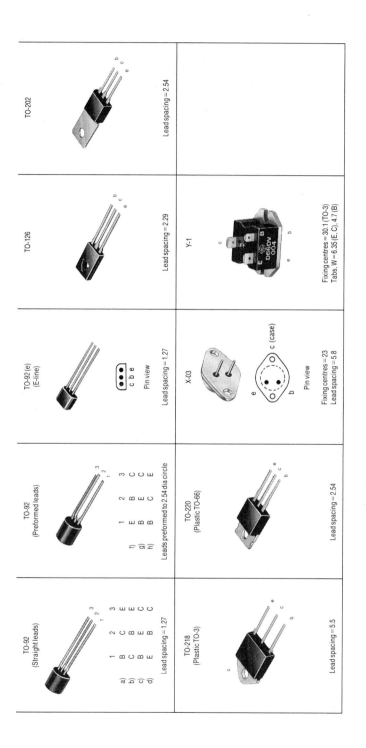

Heat Sinks

TO-220/TO-218/TO-202
Solderable tabs 2.6 dia on 25.4 fixing centres
TO-220/TO-218
175-650 13°C/W, H = 12.5, W = 34.5, D = 25
175-652 10°C/W, H = 12.5, W = 34.5, D = 38
175-654 8.6°C/W, H = 12.5, W = 34.5, D = 50
TO-202
175-655 10°C/W, H = 12.5, W = 34.5, D = 38

TO-220/TO-218
Solderable tabs 2.6 dia on 25.4 fixing centres.
Clamp spring mounting
175-649 13°C/W, H = 12.5, W = 34.5, D = 25
175-651 10°C/W, H = 12.5, W = 34.5, D = 38
175-653 8.6°C/W, H = 12.5, W = 34.5, D = 50
175-656 Retaining Clip

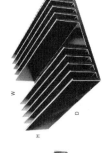

150-016 1.0°C/W, H = 37, W = 120, D = 100
150-017 0.9°C/W, H = 37, W = 120, D = 150
150-018 0.85°C/W, H = 37, W = 120, D = 200

150-019 0.50°C/W, H = 50, W = 125, D = 150
150-020 0.42°C/W, H = 50, W = 125, D = 250

150-014 2.9°C/W, H = 14, W = 93, D = 100
150-015 2.2°C/W, H = 14, W = 93, D = 150

170-756 13.9°C/W, H = 16, Dia = 39.8
(screws as shown, not supplied)

170-086 6.2°C/W, H = 16, W = 60, D = 51
170-087 3.4°C/W, H = 16, W = 60, D = 89

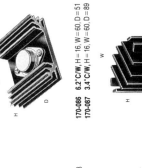

177-013 7.3°C/W, H = 25, W = 40.8, D = 26.9
Fixing centres = 30.15 (Dia. 4.1)

175-008 5.1°C/W, H = 33.3 × 46.5 sq

TO-5 62°C/W
170-754
H = 6.3, Dia = 14.3

TO-5 56°C/W
170-757
H = 15, Dia = 15

TO-5 48°C/W
170-066
H = 12.7, Dia = 14.3

TO-5 ≤30°C/W
170-064
(plastic clamp screw)
170-065
(metal clamp screw (PO1A))
H = 10.8, Dia = 23.8

TO-5 35°C/W
170-062
H = 9.5, Dia = 21

TO-18 48°C/W
170-061
H = 9.5, Dia = 18

TO-5 85°C/W
170-063
H = 7, Dia = 13

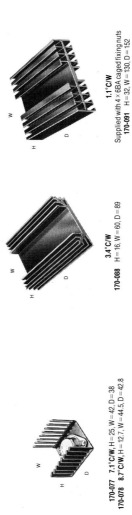

1.1°C/W
Supplied with 4 × 6BA caged fixing nuts
170-091 H = 32, W = 130, D = 152

3.4°C/W
170-088 H = 16, W = 60, D = 89

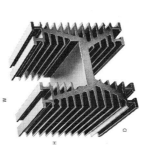

0.5°C/W H = 120, W = 120, D = 115
170-500

7.1°C/W, H = 25, W = 42, D = 38
170-077
8.7°C/W, H = 12.7, W = 44.5, D = 42.8
170-078

150-021 7°C/W
H = 25, W = 43, D = 40

170-079 14°C/W
H = 13.7, W = 44.5, D = 31.7

4°C/W and 5°C/W
To increase performance these units
may be nested together to give 2.9°C/W
170-084 4°C/W, H = 25.4 × 79.4 sq
170-085 5°C/W, H = 25.4 × 63 sq

7.1°C/W

9°C/W
To increase performance these units
may be nested with 170-084 or 170-085.
170-080 9°C/W, H = 25.4, W = 32.7, D = 41.4
170-082 7.1°C/W, H = 25 × 45.2 sq

Operational Amplifiers (Bipolar Types)

	LM301A	LM808	LM324	LM538	LM592	LM833	LM837	LT1028	NE531	LM5532	NE5534A	OP-07C
Absolute max. ratings Voltage supply range (V_{CC})	±5V to ±18V	±5V to ±18V	±1.5V to ±16V or 3V to 32V	±1.5V to ±15V or ±3V to 30V	±3V to ±8V	±5V to ±18V	±5V to ±18V	±5V to ±20V	±5V to ±22V	±3V to ±20V	±3V to ±20V	±9V to ±22V
Power dissipation	500mW	500mW	570mW	500mW	500mW	500mW	1.2W	500mW	500mW	500mW	500mW	500mW
Differential i/p volts (max)	30V	30V	32V	32V	±5V	30V	30V	±1.8V	15V	±0.5V	±0.5V	30V
Max input voltage, one input earthed	15V	15V	32V	32V	±6V	15V	15V	±12.2V	15V	13V	13V	15V
Typical ratings at 25°C with 2kΩ load												
Input offset voltage	2mV	2mV	2mV	2mV		0.3mV	0.3mV	20μV	2mV	0.5mV	0.3mV	60μV
Input offset current	3nA	0.2nA	±5nA	5nA	0.4μA	10nA	10nA	18nA	50nA	10nA	500nA	0.8nA
Input bias current	70nA	1.5nA	45nA	45nA	9μA	500nA	500nA	±30nA	400nA	200nA		±1.8nA
Input resistance	2MΩ	40MΩ			<4kΩ			20kΩ	20MΩ	300kΩ	100kΩ	33MΩ
Common mode rejection ratio	90dB	100dB	70dB	70dB	86dB	100dB	100dB	126dB	100dB	100dB	100dB	120dB
Supply voltage rejection ratio	96dB	96dB	70dB	100dB	70dB	100dB	100dB	132dB	100dB	100dB	100dB	104dB
Large signal voltage gain	104dB	110dB	100dB	100dB	52dB	110dB	110dB	150dB	96dB	100dB	100dB	112dB
Output voltage swing	±13V	±14V	±14.5V	28V	4V	±13.5V	±13.5V	±13V	±13V	±13V	±13V	±13V
Slew rate	0.4V/μs	0.2V/μs	0.5V/μs	0.5V/μs		7V/μs	10V/μs	15V/μs	35V/μs	9V/μs	13V/μs	0.17V/μs
Unity gain bandwidth	1MHz	1MHz	1MHz	1MHz	120MHz	9MHz	25MHz	75MHz	1MHz	10MHz	10MHz	0.5MHz
Full power bandwidth	10kHz	10kHz	15kHz	10kHz	20MHz	120kHz	200kHz	200kHz	500kHz	140kHz	200kHz	3.4kHz
Supply current	1.8mA	0.3mA	1mA	1.5mA	18mA	5mA	10mA	7.6mA	5.5mA	10mA	4mA	2.7mA

	OP-27G	OP-37G	OP-77G	OP-470G	μA709C	μA741C	μA747C	μA748C	1458C	3403	4136
Absolute max. ratings Voltage supply range (V_{CC})	±3V to ±22V	±4V to ±20V	±3V to ±20V	±5V to ±18V	±5V to ±18V	±5V to ±18V	±5V to ±18V	±1.25V to ±22V	±2.5V to ±18V	±18V or 2.5V to 36V	±18V
Power dissipation	658mW	500mW	500mW	500mW	250mW	500mW	800mW	500mW	500mW	500mW	800mW
Differential i/p volts (max)	0.7V	±0.7V	30V	±1V	5V	30V	30V	30V	30V	36V	30V
Max input voltage, one input earthed	±15V	±12.3V	±14V	±12V	10V	15V	15V	15V	15V	36V	15V
Typical ratings at 25°C with 2kΩ load											
Input offset voltage	30μV	30μV	50μV	400μV	2mV	1mV	1mV	1mV	1mV	2mV	0.5mV
Input offset current	12nA	12nA	0.3nA	12nA	100nA	30nA	80nA	40nA	80nA	±30nA	5nA
Input bias current	±15nA	±15nA	1.2nA	25nA	300nA	200nA	200nA	120nA	200nA	150nA	40nA
Input resistance	4MΩ	4MΩ	45MΩ	400kΩ	250kΩ	1MΩ	1MΩ	800kΩ	1MΩ		5MΩ
Common mode rejection ratio	120dB	120dB	140dB	120dB	90dB	90dB	90dB	90dB	90dB	90dB	100dB
Supply voltage rejection ratio	118dB	114dB	123dB	120dB	92dB	96dB	96dB	90dB	96dB	90dB	100dB
Large signal voltage gain	123dB	123dB	135dB	120dB	93dB	104dB	104dB	90dB	104dB	100dB	110dB
Output voltage swing	±13.5V	±13.5V	±13V	±13V	±13V	±13V	±13V	±13V	±13V	±14V	±13V
Slew rate	2.8V/μs	13.5V/μs	0.3V/μs	2V/μs	0.25V/μs	0.5V/μs	0.5V/μs	0.5V/μs	0.5V/μs	1.2V/μs	1V/μs
Unity gain bandwidth	8MHz	63MHz	0.6MHz	6MHz	5MHz	1MHz	1MHz	1MHz	1MHz	1MHz	3MHz
Full power bandwidth	34kHz	200kHz	3.4kHz	20kHz	up to 200kHz	10kHz	10kHz	10kHz	10kHz	40kHz	25kHz
Supply current	3.5mA	3mA	1.7mA	9mA	2.5mA	1.7mA	3mA	1.75mA	3mA	3mA	7mA

Operational Amplifiers (FET Types)

	CA3130E	CA3140E	CA3240E	LF347	LF351	LF353	LF411	LF412	LF441	LF442
Absolute max ratings										
Voltage supply range V_{CC}	±2.5V to ±8V or 5V to 16V	±2V to ±18V or 4V to 36V	±2V to ±18V or 4V to 36V	±5V to ±18V	±5V to ±18V	±5V to ±18V	±5V to ±18V	±5V to ±18V	±5V to ±18V	±5V to ±18V
Power dissipation	630mW	630mW	630mW	500mW	500mW	500mW	500mW	500mW	500mW	500mW
Differential input voltage (max)	±8V	±8V	±8V	±30V	±30V	±30V	±30V	±30V	±30V	±30V
Max input voltage, one input earthed	±V_{CC}	±V_{CC}	±V_{CC}	±15V	±15V	±15V	±15V	±15V	±15V	±15V
Typical ratings at 25°C										
Input offset voltage	8mV	5mV	5mV	5mV	5mV	5mV	0.8mV	1mV	1mV	1mV
Input offset current	0.5pA	0.5pA	0.5pA	25pA	25pA	25pA	25pA	25pA	5pA	5pA
Input bias current	5pA	10pA	10pA	50pA	50pA	50pA	50pA	50pA	10pA	10pA
Input resistance	1.5TΩ	1.5TΩ	1.5TΩ	1TΩ	1TΩ	1TΩ	1TΩ	1TΩ	1TΩ	1TΩ
Common mode rejection ratio	90dB	90dB	90dB	100dB	100dB	100dB	100dB	100dB	95dB	95dB
Supply voltage rejection ratio	90dB	80dB	80dB	100dB	100dB	100dB	100dB	100dB	90dB	90dB
Large signal voltage gain	110dB	100dB	100dB	100dB	100dB	100dB	106dB	106dB	100dB	100dB
Output voltage swing	13.3V (V_{CC} = 15V)	13V (V_{CC} = 15V)	13V (V_{CC} = 15V)	±13.5V	±13.5V	±13.5V	±13.5V	±13.5V	±13V	±13V
Slew rate	10V/μs	9V/μs	9V/μs	13V/μs	13V/μs	13V/μs	15V/μs	15V/μs	1V/μs	1V/μs
Unity gain bandwidth	15MHz	4.5MHz	4.5MHz	4MHz	4MHz	4MHz	4MHz	4MHz	1MHz	1MHz
Full power bandwidth	100kHz	100kHz	100kHz	100kHz	100kHz	100kHz	100kHz	100kHz	15kHz	15kHz
Supply current	2mA	4mA	8.4mA	7.2mA	1.8mA	3.6mA	1.8mA	3.6mA	150μA	400μA

	LF444	LF13741	LH0042C	TL064	TL071	TL072	TL074	TL081	TL082	TL084
Absolute max ratings										
Voltage supply range V_{CC}	±5V to ±18V	±5V to ±18V	±5V to ±22V	±2V to ±18V	±2V to ±18V	±2V to ±18V	±2V to ±18V	±2V to ±18V	±2V to ±18V	±2V to ±18V
Power dissipation	500mW	500mW	500mW	680mW	680mW	680mW	680mW	680mW	680mW	680mW
Differential input voltage (max)	±30V	±30V	±30V	±30V	±30V	±30V	±30V	±30V	±30V	±30V
Max input voltage, one input earthed	±15V	±16V	±15V	±15V	±15V	±15V	±15V	±15V	±15V	±15V
Typical ratings at 25°C										
Input offset voltage	3mV	5mV	6mV	3mV	3mV	3mV	3mV	5mV	5mV	5mV
Input offset current	5pA	10pA	2pA	5pA	5pA	5pA	5pA	5pA	5pA	5pA
Input bias current	10pA	50pA	15pA	30pA	30pA	30pA	30pA	30pA	30pA	30pA
Input resistance	1TΩ	0.5TΩ	1TΩ	1TΩ	1TΩ	1TΩ	1TΩ	1TΩ	1TΩ	1TΩ
Common mode rejection ratio	95dB	90dB	80dB	76dB	76dB	76dB	76dB	76dB	76dB	76dB
Supply voltage rejection ratio	90dB	96dB	80dB	95dB	76dB	76dB	76dB	76dB	76dB	76dB
Large signal voltage gain	100dB	100dB	100dB	75dB	106dB	106dB	106dB	106dB	106dB	106dB
Output voltage swing	±13V	±12V	±12V	±3.5V	±13.5V	±13.5V	±13.5V	±13.5V	±13.5V	±13.5V
Slew rate	1V/μs	0.5V/μs	3V/μs	3.5V/μs	13V/μs	13V/μs	13V/μs	13V/μs	13V/μs	13V/μs
Unity gain bandwidth	1MHz	1MHz	1MHz	1MHz	3MHz	3MHz	3MHz	3MHz	3MHz	3MHz
Full power bandwidth	15kHz	10kHz	40kHz	30kHz	100kHz	100kHz	100kHz	100kHz	100kHz	100kHz
Supply current	800μA	2mA	2.8mA	800μA	1.4mA	2.8mA	5.6mA	1.4mA	2.8mA	5.6mA

Thyristors

Package (Not relative sizes)	IT (AV) Amps	IGT (mA)	VGT (V)	30 to 200	400	600	800	1200
TO-92	0.8A	0.35	1.2	2N5060 (30V) 2N5061 (60V) 2N5062 (100V) 2N5064 (200V)				
TO-39		0.5	1.0	C203YY (60V)	P0102DA			
	1A	0.2	0.8	P0102AA (100V) P0102BA (200V)				
TO-92	2A	5.0	2.0	BTX18-100 (100V)	BTX18-400			
		0.2	1.0		TICP106D	TICP106M		
(A) TO-202	2.5A (A)	0.5	1.0	C106Y1 (30V) C106F1 (50V) C106A1 (100V) C106B1 (200V)	C106D1	C106M1		
(B) TL Plastic	(B)	0.2	1.5	TLS106-05 TLS106-1 TLS106-2	TLS106-4	TLS106-6		
	4A	0.2	0.8		X0402DE	X0402ME		
TO-220	5A	0.2	1.0	TIC106B (200V)	TIC106D	TIC106M	TIC106N	
		15	1.5		TXN408●			
	7.5A	15	1.5		BT151-500R (500V)	BT151-650R (650V)	BT151-800R	
		15	1.5		TXN412●			
	8A	20	1.5		TIC116D	TIC116M	TIC116N	
		25	2.0	S0810BH (200V)	S0810DH	S0810MH	S0810NH	
METRIC M5	10A	50	1.5			BTW38-600R	BTW38-1000R (1000V)	
(B) 10-32UNF2A	10A (B)	30	1.5		BTY79-400R-PHIL	BTY79-600R-PHIL	BTY79-800R-PHIL BTY79-1000R-PHIL (1000V)	
					BTY79-400R-SEM	BTY79-600R-SEM	BTY79-800R-SEM BTY79-1000R-SEM (1000V)	
(A) ¼"-28UNF2A	10A (A)	60	2.0					10RIA120

Package	Current							
TO-220	12A	20	1.5		TIC126D	TIC126M	TIC126N	
	13A	32	1.0		BT152-400R	BT152-600R	BT152-800R	
	16A	40	1.5	TYN225 (200V)	TYN425	TYN625	TYN825	
(A) 1/4"-28 UNF 2A (B) METRIC M6	16A (B)	75	1.5			BTW45-600R	BTW45-800R	
	16A (A)	60	2.0		16RIA40	16RIA60	16RIA100 (1000V)	16RIA120
	22A (A)	60	2.0	22RIA20	22RIA40	22RIA60		22RIA120
	25A (A)	60	2.0	25RIA20		25RIA60	25RIA80	25RIA120
	40A (A)	300	3.5			C147M	C147M	C147PB
	50A (A)	100	2.5		50RIA40	50RIA60	50RIA80 / 50RIA100 (1000V)	50RIA120
1/2"-20 UNF 2A	70A	100	2.5			71RIA60		71RIA120
	110A	150	3.0			ST110S06P	ST110S08P	ST110S12P
3/4"-16 UNF 2A	230A	150	3.0			ST230S06P		ST230S12P

● Insulated devices (insulation voltage 2500V rms)

Triacs

Package (Not relative sizes):

- TO-92 (Pin view: t2 g t1)
- TO-92 (Pin view: t1 t2 g)
- TO-220 (g, t2/tab, t1)
- TO-220
- TO-202 (g, t2, t1)
- TO-126
- ¼"-28 UNF 2A (g, t2)
- TO-220 — tab = t2 (non-isolated types)

IT RMS (A)	Isolation	IGT (mA) t2+ g+	IGT (mA) t2+ g-	IGT (mA) t2- g-	IGT (mA) t2- g+	VGT (V)	VDRM 400	VDRM 600
0.8A	ISOLATED	5	5	5	5	2.0	Z0105DA	
 1.5A	ISOLATED	8	8	8	10	2.5	TICP206D	TICP206M
 2A	NON-ISOLATED	5	8	10	25	2.5	TIC201D	
4A	NON-ISOLATED	35	35	35	70	1.5	BT136-500 (500V)	BT136-600
4A	NON-ISOLATED	5	5	5	10	2.0	TIC206D	TIC206M
 4A	NON-ISOLATED	5	5	5	5	2.0	Z0405DE	Z0405ME
 4A	NON-ISOLATED	5	5	5	10	2.5	2N6070A (100V) / 2N6073A	2N6075A
6A	ISOLATED	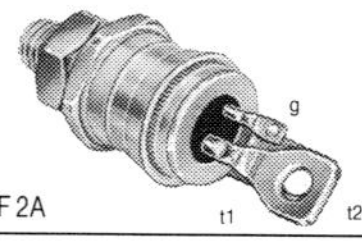 50	50	50	—	2.5	SC24D2	
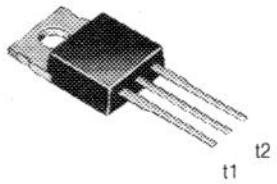 6A	NON-ISOLATED	50	50	50	—	1.5	BTB06-400BW●	BTB06-600BW●
6A	NON-ISOLATED	50	50	50	100	1.5	BTB06-400B	BTB06-600B
6A	NON-ISOLATED	5	5	5	10	2.2		TIC216M
8A	ISOLATED	50	50	50	—	1.5	BTA08-400BW●	BTA08-600BW●
8A	ISOLATED	50	50	50	100	1.5	BTA08-400B	BTA08-600B
8A	NON-ISOLATED	35	35	35	70	1.5	BT137-500 (500V)	BT137-600
8A	NON-ISOLATED	5	20	20	30	2.0	TIC225D	TIC225M
8A	NON-ISOLATED	50	50	50	—	2.0	TIC226D	TIC226M
8A	ISOLATED	50	50	50	50	2.5	TO812DJ	TO812NJ (800V)
10A	NON-ISOLATED	50	50	50	—	1.5	BTB10-400BW●	BTB10-600BW●
10A	ISOLATED	25	25	25	25	2.5	T1010DJ	T1010MJ
12A	ISOLATED	25	25	25	50	1.5		BTA12-600C
12A	ISOLATED	50	50	50	50	2.5	T1212DJ	T1212MJ
12A	NON-ISOLATED	35	35	35	70	1.5	BT138-500 (500V)	BT138-600
12A	NON-ISOLATED	50	50	50	50	2.0	TIC236D	TIC236M

Package	$I_{T(RMS)}$	Isolation									
¼"-28 UNF 2A (t1/t2/g)	15A	NON-ISOLATED	90	50	50	50	50	—	2.5	SC250D	SC250M
(A) TO-220 / (B) TO-218	16A	(A) ISOLATED	170	50	50	50	50	100	1.5	BT139-500 (500V)	BTA16-600B
		(A) NON-ISOLATED	115	35	35	35	35	70	1.5		BT139-600
		(A) NON-ISOLATED	125	50	50	50	50	—	2.0	TIC246D	TIC246M
	20A	(A) NON-ISOLATED	160	50	50	50	50	100	1.5	BTB16-400B	
		(B) NON-ISOLATED	150	50	50	50	50	—	2.0	TIC253D	TIC253M
	25A	(B) NON-ISOLATED	175	50	50	50	50	—	2.0	TIC263D	TIC263M
		(A) NON-ISOLATED	200	50	50	50	50	100	1.5	BTB24-400B	
		(A) NON-ISOLATED	180	35	35	35	35	70	1.5		BTA140-600
TO-3 FLANGE	25A	ISOLATED	230	50	50	50	50	—	2.5	SC160D	SC160M
	30A	ISOLATED	260	50	50	50	50	100	1.5	BTA25-400B	BTA25-600B
	25A	ISOLATED	230	50	50	50	50	—	2.5		
isolated	40A	ISOLATED	275	80	80	80	80	150	2.5	SC260D2	SC260M2
non-isolated		NON-ISOLATED	275	80	80	80	80	150	2.5	SC265D-2	SC265M-2
		NON-ISOLATED	265	50	50	50	50	80	2.5		SC265M-1
											2N5446
¼"-28 UNF 2A	50A	NON-ISOLATED	520	200	200	200	200	200	2.5		50AC80A (800V)
		NON-ISOLATED	520	200	200	200	200	200	2.5		50AC120A (1200V)

● Snubberless triacs do not require additional snubber networks due to their improved di/dt and dv/dt characteristics

Index